RAND NATIONAL DEFENSE RESEARCH INSTITUTE

Can We Explain Gender Differences in Officer Career Progression?

Beth J. Asch, Trey Miller, Gabriel Weinberger

Prepared for the Office of the Secretary of Defense

Approved for public release; distribution unlimited

For more information on this publication, visit www.rand.org/t/RR1288

Library of Congress Cataloging-in-Publication Data is available for this publication.
ISBN: 978-0-8330-9461-2

Published by the RAND Corporation, Santa Monica, Calif.

Cover: U.S. Navy photo by Mass Communication Specialist 3rd Class Lauren Booher/Released

Preface

In 2011, the congressionally mandated Military Leadership Diversity Commission (MLDC) concluded that two factors contributed to the underrepresentation among senior military leaders of racial and ethnic minority and female officers: lower rates of promotion than white male officers and, in the case of midlevel female officers, lower retention. Left unclear is the relative contribution of each. That is, to what extent is the lack of representation mostly because of lower retention, lower promotion rates, or both? The MLDC relied, in part, on the results of an earlier RAND study that tracked the retention and promotion of officers, using data on officer cohorts entering between 1967 and 1991 and tracking them through 1994.

Because the results of this earlier study are dated, the Office of the Secretary of Defense asked RAND to update the study, using more-recent data. The Office of the Secretary of Defense also requested that RAND provide information on what explains gender differences in the officer career pipeline. The updated analysis was conducted in the first phase of our research, summarized in Beth Asch, Trey Miller, and Alessandro Malchiodi, *A New Look at Gender and Minority Differences in Officer Career Progression in the Military* (2012). The second phase of the research is summarized in this report and addresses the question of what explains gender differences in the officer career pipeline. The analysis should be of interest to the policy community concerned about the career progression of minority and female officers and the military manpower research community.

This research was sponsored by the Office of the Under Secretary of Defense for Personnel and Readiness and conducted within the

Forces and Resources Policy Center of the RAND National Defense Research Institute, a federally funded research and development center sponsored by the Office of the Secretary of Defense, the Joint Staff, the Unified Combatant Commands, the Navy, the Marine Corps, the defense agencies, and the defense Intelligence Community.

For more information on the Forces and Resources Policy Center, see www.rand.org/nsrd/ndri/centers/frp or contact the director (contact information is provided on the web page).

Contents

Figures and Tables

Figures

Tables

Summary

An ongoing concern of personnel managers in the Department of Defense (DoD) is the lack of diversity among senior military leaders, as well as the need to improve the representation of female officers in the senior ranks. In January 2013, the Secretary of Defense lifted the restriction on the service of women in combat units, and the services had until January 2016 to provide a review of their standards and assignment policies for implementing this policy. As a result of the review, all gender-based restrictions were lifted, starting in January 2016. Despite this recent focus on gender integration, surprisingly little quantitative information is available on how the career trajectories of female and male officers differ and the factors that explain those differences, though a notable exception is an older study by Hosek et al. (2001).

In this study, we conducted a two-phase analysis to address the gap in quantitative information on differences in the career progression of officers based on gender and minority status, as well as the factors that explain these differences. In the first phase, we updated the Hosek et al. (2001) study using more-recent data. This analysis is summarized in Asch, Miller, and Malchiodi (2012). The second phase of the study is summarized in this report. The objective of this second phase is to provide quantitative information on the factors that explain gender differences in officer career progression. The analysis in both reports examines career progression as a series of retention and promotion outcomes, each conditional on having attained the proceeding outcome. Specifically, we consider ten specific career milestones: retention at O1, promotion to O2, retention at O2, promotion to O3, and so forth. The last milestone is promotion to O6.

Data and Methods

We used longitudinal data on officers provided by the Defense Manpower Data Center to track cohorts of officers entering between 1971 and 2005 over their careers, from 2000 through 2010. The data we used include information on occupation; service branch; grade; source of commission; deployments; and demography, including such variables as race, ethnicity, gender, education, marital status (e.g., dual-military status), the presence of dependents, and the ages of dependents.

We used a regression decomposition methodology, based on the well-known Blinder-Oaxaca method, to decompose gender differences in the likelihood of officers reaching each subsequent career milestone into the portion that is attributable to differences in observed characteristics and the portion that is attributable to the association between a given characteristic and the likelihood of achieving a given career milestone. We call the former part the *observed component* of the gender gap regarding the likelihood of reaching a given milestone and the latter part the *association component.* The associations we estimated capture structural factors—e.g., the factors that cause retention and promotion outcomes to differ for male and female officers with the same observed characteristics, as well as the role of self-selection and endogeneity of both observed and unobserved characteristics. Our method permits the detailed decomposition of the observed and association components into the contributions of specific observed characteristics. Thus, we are able to assess which specific characteristics, such as occupational group and age of dependents, are the most important contributors to each component of the gender gap in career progression between male and female officers.

It is important to recognize that the analysis provides descriptions of gender differences in career progression and the extent to which those differences are explained by differences in observed characteristics and the associations between characteristics and career-progression outcomes. The analysis does not assess why observed characteristics differ or why differences in factors are positively or negatively related to career progression. Furthermore, because of estimation issues related to possible selectivity biases, we must be cautious in our interpretation of the analysis and focus on general magnitudes of results rather than specific estimates. Also, because of data constraints, the analysis focuses on officer retention and promotion behavior from 2000 to 2010, a period when many officers were deployed for the wars in Iraq and Afghanistan. This limits the comparability of our results to past studies and might also result in different relationships between family status and career progression than what would happen during peacetime. Nonetheless, our study provides one of the first quantitative assessments of the factors associated with gender differences in career progression across the officer force in the Department of Defense (DoD) that explicitly considers what can and cannot be explained by observed characteristics.

Gender Differences in Career Progression and Observed Characteristics

Table S.1 shows tabulations of the gender differences in officer career progression. We find larger gender differences in the midcareer. Specifically, female officers are less likely to be promoted to O3, conditional on retention as an O2 (by 3.6 percentage points); less likely to be retained as an O3 until the O4 promotion window (by 11.8 percentage points); and less likely to be promoted to O4, conditional on retention as an O3 (by 6.0 percentage points). Beyond that point, we find relatively little difference until the O5 retention point, where female officers are less likely to stay as an O5 than male officers (by 10.9 percentage points).

Our tabulations also reveal differences in the observed characteristics of male and female officers, which, in some cases, vary across career milestones because of differences in both career progression for a given cohort and characteristics across cohorts:

Table S.1
Officers Retained or Promoted in Phase 2 Analysis: Male Versus Female Officers

Milestone	Overall (%)	Male Officers (%)	Female Officers (%)	Difference (Male Minus Female)
Retained as O1	99.9	99.9	99.8	0.1
Promotion to O2	98.1	98.2	97.2	1.0
Retained as O2	99.5	99.6	99.2	0.4
Promotion to O3	92.7	93.3	89.6	3.6
Retained as O3	82.0	83.2	71.6	11.8
Promotion to O4	83.6	84.2	78.1	6.0
Retained as O4	92.3	92.4	91.3	1.2
Promotion to O5	85.4	85.6	83.5	2.1
Retained as O5	86.2	86.2	76.3	10.9
Promotion to O6	59.2	59.3	58.0	1.3

NOTE: The last column was calculated before the percentages were rounded. The largest gender differences are highlighted.

- Female officers are less likely to be academy graduates, more likely to be in administrative occupations, and less likely to be in tactical occupations.
- Female officers are less likely to be married, more likely to be a dual-military spouse, and less likely to have dependents than male officers at each career milestone.
- Female officers have more education, are more likely to have entered a recent cohort, and are less likely to have prior enlisted service.

Decomposition of Gender Differences in Career Progression

We decomposed the gender differences in the likelihood of achieving a given milestone into observed and association components, focusing on those milestones with larger differences (bolded in Table S.1) and with relatively large observed and association components (as is the case of promotion to O6).

A key finding is that the portion of the difference attributable to variations in observed characteristics is not 100 percent; observed differences in individual and job characteristics explain some but not all of the differences in the likelihood of reaching the selected milestones (see Figure S.1).

- In the case of retention as an O3, at most the observed differences in characteristics explain about two-thirds (7.9 out of 11.8 percentage points) of the gender difference.
- In the case of retention as an O5, differences in observed characteristics only explain about one-quarter of the gender difference.

Figure S.1
Decomposition of Gender Differences in Probability of Achieving Selected Career Milestones: Explained and Unexplained Components

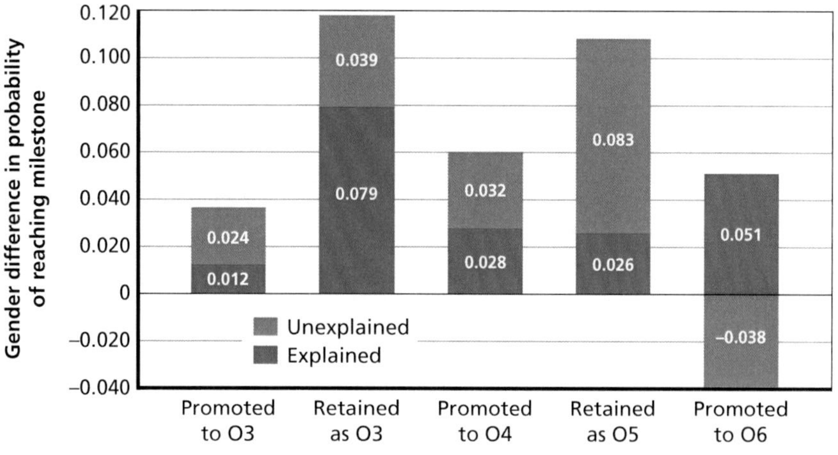

RAND *RR1288-S.1*

- In the case of promotion to O6, observed differences in characteristics contributed to a lower likelihood of promotion for female relative to male officers, but unexplained factors contributed an almost equal amount to women's higher likelihood of promotion. That is, observed factors are disadvantageous to the O6 promotion for female officers relative to male officers, but differences in the associations between factors and O6 promotion for male and female officers are advantageous and almost completely offset the disadvantageous factors.

In short, differences in observed characteristics are important contributors to gender differences in career progression but are not the only contributors.

We further decomposed the observed and association components of the gender difference in officer career progression into specific characteristics. We did this for the milestones where these components are sizable: retention as an O3, promotion to an O4, retention as an O5, and promotion to an O6. Table S.2 summarizes the results. It shows the characteristics that are the main contributors to the observed component and the association component of the gender gap in the likelihood of reaching the milestone.

Table S.2
Summary of the Main Contributors to Gender Differences in Reaching Selected Milestones

Key Milestone	Difference in Gender Gap Attributable to Variations in Observed Characteristics	Difference in Gender Gap Attributable to Variations in the Association Between Characteristic and Outcome
Retained as O3	Family status, occupation, entry year	Family status, occupation, deployment experience, prior service, entry year
Promoted to O4	Family status, occupation, entry year	Family status, race/ethnicity
Retained as O5	Occupation	Prior service, entry year
Promoted to O6	Family status, deployment experience	Family status, deployment experience, race/ethnicity

Key Findings on the Characteristics That Contributed to the Observed Component of the Gender Gap

Among the characteristics that contributed to the portion of the gender gap attributable to differences in observed characteristics, family status—including marital status and presence of children—was consistently important at the selected milestones. The lower marriage rate and lower likelihood of having dependents among female officers (relative to male officers) contributed to the gender gap at these milestones, given that being married and having dependents were both positively associated with career progression among officers. As with their civilian counterparts, better-educated women in the military postponed marriage and childbirth. Yet, also like their civilian counterparts, married men had better career outcomes (A. Miller, 2011; Buckles, 2008; Lundberg and Rose, 2000; Waite, 1995).

Occupational group and, related to occupation, cumulative months of deployment were also generally disadvantageous to female officers, in terms of contributing to the gender gap in career progression attributable to differences in observed characteristics. While our descriptive analysis does not assess the effects on career progression of restrictions on the service of women, the analysis indicates that the lower representation of female officers in tactical occupations and higher representation in administrative occupations contributed to the observed component of the gender gap in the likelihood of career progression to key milestones. Entry year was also a main contributor. Female officers were more likely to enter in recent years, and more-recent entrants had a lower likelihood of being retained and promoted in the midcareer.

Key Findings on Characteristics That Contributed to the Association Component of the Gender Gap in Career Progression

As mentioned, differences in observed characteristics were not the only contributor to gender differences in career progression. Differences in the associations between these characteristics and the probability of reaching each milestone were also important. The differences in associations captured the differences in retention and promotion outcomes for male and female officers with the same observed characteristics.

Several factors contributed to the portion of the gender gap attributable to differences in the association between characteristics and the probability of reaching the selected milestones. Although no single factor is the primary contributor, family status is a statistically significant contributing factor to the association component in all milestones except retention as an O5, where few factors are statistically significant.

Specifically, we find that male and female officers with the same family status had different probabilities of being retained as an O3 (conditional on being promoted to O3), being promoted to O4 (conditional on being retained as an O3), and being promoted to O6 (conditional on being retained as an O5). For example, having a dependent between the ages of seven and 18 is positively associated with promotion to O6 for male officers but not for female officers. This result could be due to differences in the effects of children on promotion for male and female officers or to gender differences in the self-selection of officers who become eligible for O6 promotion, where the selection mechanism depends on the presence of children.

Key Findings on the Characteristics of the Overall Gender Gap in Career Progression

In addition to considering the contribution of specific characteristics to the observed and association components, we considered the overall contribution, or sum, of the two components for each characteristic to the three milestones with the largest gender gap: retention as an O3, promotion to O4, and retention as an O5. With respect to family status, we find that

- Family status had an overall positive and large contribution to the gender gap in terms of the probability of being retained as an O3. The lower marriage rate and rate of dependents among female officers, together with differences in the association between family status and O3 retention for male and female officers, contributed to the lower O3 retention among female officers.
- Family status contributed positively, overall, to the gender gap in the probability of promotion to O4 but had no statistically sig-

nificant overall contribution to gender differences in retention as an O5.

In short, family status tended to be disadvantageous to female officers in terms of contributing positively the gender gap in O3 retention and O4 promotion.

In contrast, occupational group differences were advantageous, overall, to female officers for being retained as an O3; although female officers were less likely to be in tactical occupations, which were more likely to be retained, the association between O3 retention and being in a tactical occupation was stronger for female officers. Thus, the negative contribution of the association component outweighed the positive contribution of the observed component, so the overall effect was advantageous for female officers. The opposite is the case for O5 retention; occupational group differences were a positive contributor overall to the gender gap in O5 retention.

Entry year was also a notable contributor to the gender gap in reaching key milestones. Female officers were more represented in recent entry cohorts, while female officers who entered in these recent cohorts had lower retention than male officers. Both the association component and observed component of the entry cohort variables are disadvantageous to female officers.

We find that being a dual-military spouse had little or no role in contributing to the explained and unexplained gender differences in officer career progression. This finding is surprising given the attention that issues and challenges facing dual-military couples have received from the research and policy communities (L. Miller et al., 2011; Smith, 2010; Moini, Zellman, and Gates, 2006; Steinberg, Harris, and Scarville, 1993; Teplitzky, 1988).

Our study does not necessarily imply that dual-military couples do not face challenges with such issues as colocation and finding adequate and dependable childcare, as past research has demonstrated. However, the study does suggest that such factors might not translate to material differences in the career progression of dual-military spouses. It could be that civilian spouses of officers tend to have high-stress, professional careers that could also contribute to lower retention rates for their officer spouses. Alternatively, it could be that dual-military spouses exhibit higher attachment to their military careers, which might counterbalance the negative influence of those challenges.

Policy Implications

The results of our study have several policy implications, though it should be remembered that our analysis is descriptive and does not directly evaluate existing policies to reduce the gender gap in officer career progression. Given our findings about the role of characteristics in contributing to the gender gap, and specifically of family status and occupational group, policies that reduce differences in these characteristics (such as the lifting of restrictions on the service of women) are likely to contribute to the narrowing of the gender gap.

However, the analysis also indicates that these policies are unlikely to fully eliminate the gender gap, given the role of differences in the association between factors and career milestones. The associations we estimated captured structural factors and the role of self-selection and endogeneity of both observed and unobserved characteristics. With respect to unobserved characteristics, gender differences in taste for the military and performance could be important. Policies that improve attitudes toward service (such as those that address sexual harassment and assault), for example, could have a role insofar as they address the unexplained portion of the gender gap. We find that multiple factors contributed to the association portion of the gender gap, though family factors are among these factors. Thus, our analysis suggests that policies aimed at targeting work-family balance are likely to reduce the gender gap, given our findings on the important role of these factors in contributing to the gender gap. Finally, we find that dual-military officers exhibited similar patterns of career progression as other officers. It could be that the programs and policies that the services have developed to address issues among dual-military spouses, such as prioritizing the colocation of spouse duty, have helped.

Acknowledgments

We are grateful to the Defense Manpower Data for providing the Proxy-PERSTEMPO data and DEERS-PITE data used in our analysis. We benefited from the comments we received from the Office of Accession Policy (AP) within the Office of the Under Secretary of Defense for Personnel and Readiness (Military Personnel Policy). We would especially like to thank Dennis Drogo, our project monitor, within AP. We would also like to thank Arthur Bullock at RAND, who provided excellent programming support to our project. We are grateful to the comments from Larry Hanser and Susan Hosek, who provided a technical review of the report, and we thank them for their input.

CHAPTER ONE

Introduction

This report is the second of two that provide information on the career progression of officers in the military and differences by race and ethnicity and gender. Both reports were motivated by an ongoing concern within the Department of Defense (DoD) about the diversity of the military's leadership, especially in the more-senior officer corps. While diversity has increased historically (Lim et al., 2014), minority and female officers are less likely to be in the senior-officer ranks (O4 through general and flag officer ranks) than in the junior-officer ranks. Table 1.1 shows that in fiscal year (FY) 2013, the most recent year for which DoD has published data, female officers composed 19.4 percent of junior officers in the grades of O1 to O3, but 13.8 percent in grades O4 to O6 and 7.7 percent of general and flag officers. While the percentages differ, the pattern is similar for racial minorities and Hispanics. Thus, officer diversity remains an ongoing concern.

Another motivation for the analysis was the lifting of the restriction on the service of women in combat units. The services had until January 2016 to provide a review of their standards and assignment policies for implementing this change. As a result of the review, all gender-based restrictions were lifted, beginning in January 2016. Despite this recent focus on gender integration, surprisingly little quantitative information is available on how the career trajectories of female and male officers differ and the factors that explain those differences, though a notable exception is an older study by Hosek et al. (2001).

Understanding the underrepresentation of minority and female officers in the senior ranks requires an understanding of differences in

Table 1.1
Active Component Officer Corps: Gender, Race, and Ethnicity Status, FY 2013

Status	Accessions (%)	O1 to O3 (%)	O4 to O6 (%)	General and Flag Officers (%)	All Officers (%)
Female	22.8	19.4	13.8	7.7	17.1
White	75.9	77.2	80.5	90.4	78.6
Black	7.5	8.3	9.0	6.6	8.6
Asian	5.4	5.1	3.8	1.8	4.6
Other, two or more, unknown	11.3	9.4	6.7	1.2	8.3
Hispanic	6.2	6.0	5.3	1.4	5.7

SOURCE: DoD., 2015, Tables B-23, B-25, B-38, and B-39.

their career progression and the factors that affect those differences. *Career progression* refers to the process by which individuals become an officer, pursue their military careers, and advance through the ranks. Differences in career progression might be due to a number of factors, including entry source and qualifications, occupation and job assignment, retention behavior, promotion selection criteria, and performance.

Study Objective and First-Phase Report

In our first report (Asch, Miller, and Malchiodi, 2012), we focused on two aspects of the career progression process for officers: promotion and retention. That report analyzed gender and minority differences in the attainment of successive promotion and retention milestones of entry cohorts of officers. It also analyzed differences in career progression among female officers in partially closed versus open occupations. The report provided a descriptive analysis of the extent to which lower promotion, lower retention, or both factors contributed to the lower representation of minorities and female officers among senior military leaders. The report used data on entering officer cohorts, tracking their progression to each retention and promotion career milestone, O1 through O6, until 2010. That analysis also estimated differences between the career progression of female officers in occupations partially closed to women and female career progression in open occupations. The results were inputs to the Secretary of Defense's 2012 report to Congress on restrictions on the service of women (Office of the Under Secretary of Defense, Personnel and Readiness, 2012).

A key finding of the first report was that career progression does indeed differ for minority and female officers relative to white men. Minority male officers are more likely to be retained at each milestone but also less likely to be promoted, conditional on retention. In the early career, the lower promotion rate was offset by higher retention, so the likelihood that a minority male officer reached O4 was higher than for white male officers. However, in the field grades of O4 to O6, the lower promotion offset the higher retention. In particular, black male officers had a 19.5-percent likelihood of reaching O6, conditional on having reached O4, compared with 23.6 percent of white male officers. As a result of lower promotion rates, minority male officers were less likely to be among the pool of personnel from which senior general and flag officers are chosen, especially to the O4 and O5 milestones, despite higher retention.

For female officers, the report found some differences by race and ethnicity. However, the report generally found that female officers made slower progress in the early career to O4 than white male officers did, largely owing to lower promotion rates to O3 and O4, as well as lower retention following the O3 promotion. Female officers also made slower progress to O6, conditional on having reached O4, largely owing to lower retention after reaching O5. Thus, female officers were less likely to be part of the pool from which general and flag officers are selected, generally because of significantly lower retention at the O3 and O5 levels, as well as lower promotion to the O4 level. Finally, the report found that, on net, relative to men, women in partially closed occupations were as likely as women in open occupations to reach O6, conditional on having reached O4.

The earlier report was a descriptive analysis intended to provide updated and quantitative information on how career progression differs and at which points in careers those differences arise. The report did not attempt to examine the factors that are associated with those differences. The second half of our study focuses on this question.

This report provides quantitative analysis on some of the factors associated with differences in career progression, focusing specifically on female officers relative to male officers. The study provides information on the relative contribution of different factors toward explaining differences in career progression, particularly information on the relative importance of differences in the incidence of different factors versus differences in how a given factor is associated with career progression. For example, officers who are retained are more likely to be married, but female officers have a lower incidence of marriage over their careers. However, being married is positively associated with career progression at some stages for male officers but not for female officers.

Our analysis is designed to provide information on such questions as: To what extent is lower retention among female officers at each career milestone because of their lower incidence of marriage relative to male officers, to other observed differences between male and female officers, or to differences in retention even when men and women have the same observed characteristics? Similarly, the analysis is designed to address to the extent to which observed differences in promotion to each grade are attributable to observed differences in the characteristics of male and female officers versus differences in promotion, even when the observed characteristics are the same. We consider individual characteristics—focusing on family status and marital status, as well as presence and age of dependents—and job characteristics, including occupational area.

The role of occupational area is particularly salient. In January 2013, the combat exclusion policy was lifted on the service of women in combat positions in the U.S. military. Part of the argument made by the 2011 report of the Military Leadership Diversity Commission, as well as the 2012 Office of the Under Secretary of Defense, Personnel and Readiness, report of restrictions on the service of women in the armed services, both in support of eliminating the combat exclusion, was that such the exclusion policy was an institutional barrier to career advancement. Female officers are less likely to be in occupational areas that are more likely to lead to future promotion. That is, senior leaders are disproportionately in occupations from which women faced restrictions in serving. While our analysis does not specifically focus on the effects of opening occupations on the career progression of female officers, it does provide information on the relative role of occupational group versus other factors in explaining differences in the career progression of female and male officers.

Family factors are also highly relevant. The Navy and Air Force are testing programs that permit service members to take a sabbatical or "career intermission" (Navy Personnel Command, n.d.; Losey, 2015). Part of the objective of these programs is to allow members to take time off to start or expand their families and more broadly help service members balance service and family life. Insofar as such family factors contribute to lower retention, one of the intentions of a sabbatical program is to improve retention and satisfaction with service. Our analysis of career progression examines the extent to which such factors explain differences in career progression for female versus male officers and the relative importance of these factors versus other factors, such as occupational area.

Past Literature

Our analysis contributes to a sparse but growing body of research exploring factors that might contribute to differences in the career progression of male and female officers. Broadly speaking, this literature suggests that family- and job-related characteristics play a role in officer career progression. Here we focus on family-related characteristics, since these factors are additions to our phase 1 analysis (the phase 1 analysis included job-related characteristics).

Qualitative studies have found that female service members indicate that time separated from family and work or life conflicts are their top reasons for leaving the military (Jones, 1997; Steinberg, Harris, and

Scarville, 1993). An unpublished quantitative analysis from RAND found lower continuation rates for female officers with a child younger than one at home. Another quantitative analysis indicated that being married and having dependents is positively related to officer retention for men but has no statistically significant relationship with retention for female officers (Kraus et al., 2013). Female officers cited difficulty in finding adequate childcare, particularly during times of deployment (Long, 2010; Smith, 2010; Steinberg, Harris, and Scarville, 1993; Teplitzky, 1988). Interviews and focus groups of female and male officers, summarized in Hosek et al. (2001), revealed that perceived limited occupational roles, concerns about harassment, and competing family obligations were the main reasons cited for why female officers separate from the military at a substantially greater rate than men. Lim et al. (2014) considered gender and minority differences in Air Force officer eligibility, accessions, retention, and promotion. With respect to gender differences and Air Force officer retention, the study found that the observed lower retention rates of female officers are partially but not fully explained by differences in observed factors, including marital status and presence of children, especially later in the officer career. However, the study found few differences in the promotion rates of male and female Air Force officers after controlling for observed characteristics, including these family-related factors and occupation.

Female officers are more likely to be married to another service member than are male officers, and available research suggests that being a joint spouse can affect career progression. Research suggests that dual-career officers can have difficulty maintaining a joint domicile, and such difficulties are an important contributor to the decision to leave (Smith, 2010; Steinberg, Harris, and Scarville, 1993; Teplitzky, 1988). Long (2010) found that deployments have a negative relationship with retention intentions, and the negative effect is larger for dual-career members than for other groups. Moini, Zellman, and Gates (2006) found that dual-military parents were substantially more likely than single parents to state in a survey that they would consider leaving because of childcare concerns. Laura Miller et al. (2011) found a similar result in the context of a survey they conducted of Air Force families. That study found that obtaining childcare during work or school hours was cited as a problem more frequently for dual-military spouses than for civilian spouses.

That said, a major limitation of the existing literature is that, aside from being rather sparse, no previous study of gender differences in career progression decomposed the differences into the portion explainable by differences in individual and job characteristics and structural factors, such as the promotion process of the underlying retention decisions that relate to how those differences affect career progression. While observed factors might affect gender differences in career progression, such as having children, the role of these factors might be quite small relative to structural or unobserved factors.

Another drawback of the existing literature is that past studies usually focused on a specific service, or occupation, as opposed to the officer corps as a whole. In addition, some of the studies are dated (from before 2000). While the qualitative studies provide rich detail on relevant factors, they focus on retention intentions, not actual retention behavior. Furthermore, interview and focus group analysis cannot be generalized to the officer corps. Many of the quantitative studies are more recent but are often largely descriptive, focusing on a specific service or occupational area.

Our research differs from earlier analyses in a number of other ways. First, it considers the officer community as a whole and does not focus on officers in a specific service branch or community. This means that we are not able to provide narrowly focused results for a specific area of the military, but it also means that we provide a broad overview from a DoD perspective. Second, the analysis examines career progression as a series of retention and promotion outcomes, each conditional of the proceeding outcome. We consider ten specific career milestones: retention at O1, promotion to O2, retention at O2, promotion to O3, and so forth. The last milestone is promotion to O6. The advantage of this approach is that by considering the progression of retention and promotion outcomes separately, we can assess the extent to which a given factor is important, because of its relationship to retention (conditional on promotion), versus its relationship to promotion, given retention and the specific promotion or retention milestones that are important. Studies that only consider retention at a given career decision point, without conditioning the sequence of retention and promotion outcomes that led to the member being at that decision point, are unable to make such an assessment.

Finally, our analysis explicitly considers to what degree observed differences in the job and individual characteristics of female and male officers explain observed differences in their career progression and the degree to which career progression differences are because of structural or unobserved factors. To the extent that differences in observables explain gender differences in career progression, our study further decomposes observed differences into the contribution of individual characteristics. The decomposition is important because policy to reduce gender differences and improve the diversity of the senior officer force tend to focus more on reducing differences in observed characteristics and less on reducing differences in how those characteristics are related to career progression. Insofar as the latter is important, policies focused on addressing differences in observed characteristics will not be fully effective in reducing gender differences.

Approach

The approach we used built on the data and analysis in the first report. We use Defense Manpower Data Center (DMDC) Proxy-PERSTEMPO data on entering cohorts of officers, tracking their careers through 2010, supplemented with data from the DMDC Defense Enrollment Eligibility Reporting System Point-in-Time Extract (DEERS PITE) file on family-related factors. For each of the ten outcomes, we estimated a linear probability regression model that produced estimates of the association between individual and job-specific factors and the outcome of interest. We used these models to estimate how much of the observed difference in the outcome between male and female officers is attributable to the set of observed factors we considered and how much is due to unobserved factors or to the effects of both observed and unobserved factors on the outcome. That is, we quantified how much of the differences we observed are explainable by the fact that male and female officers have a differing set of individual and job characteristics and how much is unexplainable because the male and female officers with the same characteristic have differing outcomes. We then further decomposed how much of the differences in observables and differences in associations is attributable to each specific factor, such as marital status.

It is important to recognize that the analysis provides descriptions of how career progression differs by gender and the extent to which those differences are attributable to factors we can observe in our data. The analysis does not explain why the differences in factors occur, nor does it ascertain whether differences in factors cause differences in career progression. We did not control for every relevant factor that could affect differences in career progression. Because the analyses are purely descriptive, readers should take care not to attribute a causal explanation for the results. Nonetheless, the report provides one of the first broad overviews, from a DoD perspective, of the factors associated with differences in officer career progression and considering each milestone in that progression, which is conditional on reaching the previous milestone.

Organization of the Report

Chapter Two provides an overview of the data and methods. We present our descriptive statistics and an overview of the regression results in Chapter Three. Chapter Four presents the decomposition of the gender differences in officer career milestones into explainable and unexplainable components. In Chapter Five, we discuss policy implications and conclusions. The report also has two appendices. Appendix A provides details about the decomposition methodology we use, and Appendix B shows more-detailed results of our analysis.

CHAPTER TWO

Overview of Data and Methods

Because the data and analysis used in this report build on the data and analytic approach used in the first phase of our study, we begin with a brief overview of the data used in the phase 1 analysis and then discuss how we supplemented the data, as well as other data changes for phase 2. The chapter then describes the method we used to decompose gender differences in the probability of reaching each officer career milestone in the explainable and unexplainable parts and further decompose the explainable part into the parts attributable to observed characteristics. This method is known as the Blinder-Oaxaca decomposition, and we describe the method in this chapter and its application to gender differences in officer career pipelines.

Overview of Data

Description of Data Used in the Phase 1 Analysis

Our analysis extends a rich longitudinal data set that we created for the phase 1 part of our study. The data set tracks cohorts of officers from January 1988 (or time of entry) until December 2010 (or time of separation). We described in detail how we built this file in Asch, Miller, and Malchiado (2012), but we provide an overview of the process here.

The phase 1 data file was built from the Proxy-PERSTEMPO, a file that was maintained by DMDC until 2010 and contains longitudinal administrative records on active-duty personnel, from January 1988 through September 2010. For officers, the data include service, occupation (using the DoD occupational coding), grade, months of service before attaining current grade, source of commission, date of entry and date of commissioning, demographics (including race, ethnicity, gender, marital status, and education), prior enlisted service, and indicators of deployment based on receipt of two deployment-related pays (family separation pay and hostile fire pay).

Using these data, we were able to ascertain for each officer in the data their entry path in terms of commissioning source and prior service, their promotion path, and whether and when they left active duty. We used this information to construct the career progression of each officer in terms of retention and promotion, as described below.

We excluded officers who did not enter the officer ranks at the grade of O1, and we also excluded officers in professional occupations, based on their occupation coding, such as medical, legal, and religious career fields.[1] These officers were put into a separate competitive category for promotion, so their career paths were not consistent with the other officers we studied.

We measured career progression as a series of retention and promotion milestones, each conditional on its predecessor. Retention is conditional on achieving the previous grade (except O1, where it is conditional on officer commissioning), and retention is measured up to the point of eligibility for the next promotion (e.g., the promotion window). For example, retention as an O3 is measured for those who achieved O2 and remained on active duty until the beginning of the promotion window to O3.

Determining these career milestones in DMDC data, including the Proxy-PERSTEMPO data, is challenging because the data do not indicate who was considered eligible for promotion. We identified a three-year promotion eligibility window for each grade, cohort, and service based on observed promotions in the data. In general, for each grade, cohort, and service, we identified the six-month period when at least 95 percent of all promotions occurred. This six-month period was then designated as the center of the promotion window for that grade,

[1] For pre-1988 cohorts, we did not observe entry as O1, so we matched officers to a cohort based on the first observed promotion. These cohorts might include officers who entered after O1.

cohort, and service, and we added 15 months prior to this period and 15 months after this period, for a total of 36 months. Given these promotion eligibility windows, a promotion occurs if an eligible officer achieves promotion to the next grade during that window. If the officer was promoted after the window, he or she is considered not promoted.

After defining the promotion window, we defined the retention milestones. Retention is defined as staying until at least the first month of the promotion window. For example, retention as an O3 is defined as including all officers in an entry cohort and service who achieved O3 and who stayed in service at least until the first month of the promotion eligibility window for O4 for that cohort or service. A limitation of this approach is that some officers will choose to voluntarily separate during promotion windows, even when they would have a relatively high chance of promotion. Our approach wrongly classified these officers as not being promoted, when they should have been classified as failing to retain. Nevertheless, we believe that our approach, which has been used in past studies, accurately captures the majority of promotion and retention outcomes (Hosek et al., 2001; Asch, Miller, and Malchiodi, 2012).

Addition of DEERS Data

To extend the analyses reported in Asch, Miller, and Malchiodi (2012) for phase 2 of our study, we merged detailed individual-level records from DEERS records that are maintained by DMDC. DEERS contains information on all service-connected individuals eligible for military benefits, such as the TRICARE health benefit. Because service members use DEERS to register their dependents as beneficiaries of military benefits, DEERS allowed us to collect longitudinal and detailed information on family characteristics, including marriage and number and ages of dependents, for a large subset of the officers in the phase 1 file. More specifically, our source file for DEERS data has monthly point-in-time extracts for all service-connected individuals from January 2000 to December 2014, allowing us to match family-related variables to all observations in our phase 1 file, from January 2000 to September 2010.

Because family-related variables change over time, as officers get married or divorced and gain or lose young dependents, we matched the records in the phase 1 file to the proper observations in the DEERS. That is, we ensured that time-varying characteristics were properly matched in terms of timing with the appropriate career milestone. To do this, we captured DEERS data for each officer in the phase 1 analysis during the month in which he or she achieved each career milestone that occurred between January 2000 and September 2010. For example, for an officer who entered as an O1 in 2001 and was promoted to O3, we captured family-related variables from DEERS during the months in which the officer entered, reached the promotion window to O2, was promoted to O2, reached the promotion window to O3, and was promoted to O3. Thus, we measured time-varying characteristics at the time the milestone was reached.

One challenge with time-varying characteristics is that we had to measure family characteristics for those who leave. For example, some officers might leave before reaching the O4 promotion window, so they are not retained as an O3. We had to decide when to measure their family characteristics while modeling the probability of being retained as an O3. One approach to measuring family characteristics is to measure them at the end of the previous milestone. In our example, this would mean measuring family characteristics when personnel were promoted to O3. The problem with this approach is that there can be a number of years between promotion windows. Within those years, an officer's marital and dependents status can change considerably, so we would measure family characteristics with considerable error if we used family characteristics as of the previous milestone. Another problem is that this approach would further limit our sample sizes, since any officer in the analytic file would need to have had at least one promotion after 2000 to observe his or her family characteristics at the previous milestone. Instead, we used a second approach. Specifically, we measured characteristics of those who left at the time of exit and characteristics of those who stayed at the time of the milestone. Thus, an O3 who left before reaching the O4 promotion window would have his or her family characteristics measured at exit, rather than at the time of the O3 promotion, while an individual who stays would have his or her characteristics measured at the beginning of the O4 promotion window. A disadvantage of this second approach is that those who leave will have their characteristics measured before those who stay.

As in Asch, Miller, and Malchiodi (2012), we only analyzed results for complete promotion windows. In other words, if the end of the data in 2010 occurred prior to the end of the promotion eligibility window or the end of a retention window, we excluded the observation from the analysis. In short, we were able to analyze results for all post-2000 officer career milestones from entry through the last full promotion window that ended prior to September 2010. Table 2.1 shows the officer cohorts that are included in analyses for each career milestone in both the phase 1 and the current analyses.

Also, as with the phase 1 analysis, the phase 2 analysis drew on more-recent officer cohorts to study early career milestones. Older officer cohorts were used to study late career milestones. For example, the phase 2 analysis of promotion to O2 drew on the 1998–2005 officer cohorts, whereas analyses of promotion to O6 drew on the 1976–1988 cohorts. It is also important to note that because of the sample restric-

Table 2.1
Career Progression Milestones and Cohorts Used in Phase 1 and 2 Analyses

Career Milestone	Entering Cohorts Used (Phase 1)	Entering Cohorts Used (Phase 2)
Retained as O1	1988–2002	1998–2005
Promoted to O2	1988–2002	1998–2005
Retained as O2	1986–2002	1996–2005
Promoted to O3	1986–2002	1996–2005
Retained as O3	1983–2002	1988–2002
Promoted to O4	1983–1999	1988–2002
Retained as O4	1977–1993	1984–1993
Promoted to O5	1977–1993	1984–1993
Retained as O5	1971–1991	1976–1988
Promoted to O6	1971–1991	1976-1988

tions imposed by the availability of DEERS data, the phase 2 analyses for each milestone drew on later officer cohorts than the phase 1 analysis. For example, the earliest cohort used to study promotion to O6 is the 1971 cohort for the phase 1 analysis, versus the 1976 cohort in the phase 2 analysis.

Table 2.1 shows that, in the phase 1 analysis, we only included observations through 2002, whereas the phase 2 analysis drew on observations through 2010. We limited the phase 1 sample to pre-2003 observations because the federal government changed the way that race and ethnicity was recorded beginning January 1, 2003. Since the phase 1 analysis focused on differences in career progression by race and ethnicity, as well as by gender, we were unable to include observations after 2002. The phase 2 analysis focused on differences by gender, so we included observations after 2002.

A clear distinction between the phase 2 analysis and the phase 1 analysis, as well as the Hosek et al. (2001) study, is that the phase 2 analysis focused on officer retention and promotion behavior from 2000 to 2010, when many officers were deployed for the wars in Iraq and Afghanistan. It is also possible that career-progression decisions during wartime might relate quite differently to family status. For example, officers might choose to delay marriage and childbearing to serve their country during a time of great need. Likewise, officers might be more willing to accept family-related hardships during wartime than during peacetime. While our focus on the wartime cohorts limits the comparability of our results to past studies, and might paint a different picture, the choice was made because of the data constraints described above. Nevertheless, the choice to focus on later cohorts is not without merit, as it allows us to paint a descriptive picture of the relationship between family status and officer career progression during the recent past, a period when the military has been focused on improving the gender diversity of the officer corps.

Table 2.2 shows a comparison of the percentage of officers reaching each retention and promotion in the current analysis versus the phase 1 analysis. The cohorts included in the phase 2 analysis have slightly higher promotion and retention rates at all career milestones,

Table 2.2.
Phase 2 Analysis and Phase 1 Analysis of Officers Retained or Promoted

Milestone	Phase 1 Analysis[a]	Phase 2 Analysis
	Promotion (%)	
O1 to O2	97.3	98.0
O2 to O3	90.8	92.7
O3 to O4	76.1	83.5
O4 to O5	74.6	84.5
O5 to O6	46.4	58.7
	Retention (%)	
O1	99.8	99.9
O2	99.3	99.5
O3	70.3	82.0
O4	88.5	91.5
O5	80.3	85.8

SOURCE: The updated analysis is based on the authors' calculations. The phase 1 analysis results are from Asch, Miller, and Malchiodi (2012).

[a] The phase 1 analysis did not include the data elements from DEERS. The addition of these elements meant that we only had usable data after 2000.

which might reflect differing promotion processes and retention behavior for cohorts entering after 2002; the phase 1 study did not include officer cohorts entering after 2002. Nevertheless, the general patterns are similar across data sets—high retention and promotion at O1 and O2, lower retention rates at O3 and O5, and lower promotion rates to O6. Moreover, the observed differences between male and female officers are qualitatively similar across both data sets. Chapter Four will show the differences in the percentages of reaching each milestone for male and female officers; the largest differences occur at the same milestones in the new analysis as in the phase 1 analysis.

Characteristics Included in the Analysis

The resulting analytic file includes the demographic and job-related factors from the phase 1 analysis but also adds the family-related characteristics we drew from the DEERS. More specifically, we included the following covariates from Asch, Miller, and Malchiodi (2012) in our models: service, source of commission, prior enlisted service, occupation group, deployment experience, and education.

The addition of DEERS allowed us to also include additional covariates: marital status, joint marriage to another service member, numbers and ages of dependents, and new dependents (e.g., recent birth or adoption).[2] These variables add richness to our analysis by permitting us to consider family status in describing career pipeline differences between male and female officers; past studies (reviewed in Chapter One) have pointed to the role of family factors in retention and promotion outcomes for female officers. That said, it is important to recognize that the DEERS data only allowed us to observe family status while officers were in the military. Female officers who deferred marriage or childbearing because of concern about the effects on their early career outcomes or who left the military to get married or have children will appear as unmarried or without children in our analysis. Consequently, we could not quantitatively assess the causal effects of family status on career outcome differences, as we discuss later in this chapter.

Finally, we included indicator variables of entry year. As shown in Table 2.1, the entry cohort calendar years differ for each milestone. To facilitate reporting of the results, we included indicators of year of entry relative to the first entry cohort. For example, our model of retention as an O3 includes entry cohorts from 1988 to 2002. For this milestone,

---

[2] The DEERS data allow us to identify single parents in our data and the children of single- versus non-single parents. In our initially regression analysis, we included separate covariates for the dependents for single versus non-single parents, but we found that the samples were not large enough and the variables were generally not statistically significant. We therefore considered marital status as married versus unmarried and separately considered the presence of children of difference age groups. Finally, we conducted exploratory analysis of whether spouses and children were collocated but found that it was not always easy to identify collocation with the DEERS data. Future analysis should consider these topics in more depth.

the first entry cohort represents 1988, the second entry cohort represents 1989, and so on, through the 14th entry cohort, which represents 2002. On the other hand, our model of promotion to O6 includes entry cohorts from 1976 to 1988, so entry cohort 1 for this milestone represents 1976, while entry cohort 12 represents 1988.

It is important to recognize that we did not include all factors that can influence career progression, such as performance, behavior, and physical fitness, because we lack data on these factors. Furthermore, while we included broad occupational group categories, we did not control for individual occupation within each group in our analysis. Thus, to the extent that there are promotion and retention differences across more narrowly defined occupations within an occupational group, our occupational variables will not account for these differences in estimating gender differences and the role of observable factors.

In sum, we use a number of variables to capture differences in observable characteristics between male and female officers and the association of these characteristics with career outcome variables. The variables we used are

- marital status
- joint-duty status
- number of children
- ages of children
- entry cohort year
- service branch
- source of commission
- prior enlisted service
- occupational group
- cumulative months of deployment
- education
- race
- ethnicity.

Approach

This second stage of our study focuses on the question of what factors account for observed differences in the achievement of each career milestone between male and female officers. Observed differences in outcomes between male and female officers can be attributed to

- differences in the job and individual characteristics that influence the achievement of each milestone
- differences in the effect of these characteristics on outcomes.

The first source of difference focuses on *explainable* factors—the differences in the observed characteristics themselves. It is important to recognize that these characteristics can include both those observed in the data, such as occupation, and those unobserved, such as performance and tastes and attitudes toward service. The second source of difference focuses on *structural* factors—differences in how the same characteristics for male and female officers affect outcomes. For example, the occupations that male and female officers enter differ (as will be shown in Chapter Four). To what extent are differences between male and female officers in the probability of being retained and being promoted over an officer career attributable to these occupational differences? To what extent is the likelihood of being promoted and being retained different for male and female officers in the same occupational group?

Oaxaca (1973) and Blinder (1973) developed a method to decompose differences in outcomes between their explainable and structural components. The method has been widely used in the literature—for example, to decompose differences in male and female pay into explainable and structural components—and the method has been refined over time.[3]

---

[3] This literature uses different terminology for the two components. Some studies refer to them as the *explainable* and *unexplainable components*, while others refer to them as the *observable* and *structural components*. We refer to them as the *explainable* and the *association components*.

We use the Blinder-Oaxaca method to decompose the observed differences in milestones for male and female officers. Neumark (1988), Cotton (1988), and Fortin (2007) further extended the methodology, and we used their extended methodology. Appendix A presents the methodology we used in this analysis, drawing on Jann (2008) and Fortin, Lemieux, and Firpo (2011). The appendix also discusses methodological challenges and the implications for the interpretation of the results. We provide a brief summary of the methods here.

The Blinder-Oaxaca method requires a regression analysis that provides estimates of the relationship between each observed characteristics and the outcome of interest. In our analysis, we considered ten milestones, so there were ten regression analyses. We then used these regressions to decompose the observed mean difference in each outcome between male and female officers into two parts:

1. the part attributable to mean differences in observed characteristics
2. the part attributable to differences in the association between that characteristic and the outcome.

The second part shows the difference in mean outcome for male and female officers with the same observed characteristics.

In performing the analysis, there are three issues we address, as we describe in more detail in Appendix A. The first concerns the choice of benchmark for evaluating differences in outcomes. The choice of benchmark is arbitrary and could lead to differing results. We followed the recent literature and pooled information on both male and female officers in a single regression analysis for each outcome and used the pool regression estimates in our decomposition. The second issue relates to how to specify categorical variables in the regression analysis given that the choice of omitted categories can affect the Blinder-Oaxaca decomposition. Again, we followed the literature and transformed the categorical variables so that the decomposition was independent of the choice of omitted category. The transformation means that we included a category that is traditionally omitted, and we have regression coefficients for each category, as shown in Table B.2, where we show the regression results.

The final issue relates to why the second part of our two-part Blinder-Oaxaca composition only shows the different associations and not the difference in the effects of characteristics on outcomes. That is, the second part does not show gender differences in the effects of marriage, children, and other characteristics on outcomes, only gender differences in the association between these variables and outcomes. We are unable to give a causal interpretation of the results because of the potential influence of self-selection and endogeneity. Officers at a given career milestone are those who made retention decisions or were selected for promotion in the past and who possibly made those decisions based on future promotion and retention prospects. This selectivity effect could differ for men and women and be based on unobserved characteristics and result in biased regression coefficient estimates of the causal effect of an observed characteristic on a specific outcome.

Observed characteristics, especially those related to family formation, could be endogenous. Individuals might choose the timing of marriage and childbearing based on their expectations about current and future promotions, as well as retention plans. Officers might marry and have children and then decide whether to stay or leave. Alternatively, they might leave because they plan to marry or have children or they might defer marriage or childbearing until after completing an obligation or achieving a promotion. As with the case of self-selection, not accounting for this endogeneity could result in biased coefficient estimates. Similarly, characteristics that we do not observe might be correlated with observed characteristics, and the correlation might differ for male and female officers. This could also bias the coefficient estimates of the causal effects of observed characteristics with an outcome.

Because of this third issue, we did not give a causal interpretation to the decomposition of the second part. In the results that we present, we refer to the first part as the *observable* part and the second part as the *association* part.

CHAPTER THREE

Descriptive Statistics and Regression Results

This chapter presents means of the characteristics in our analysis and the regression results. We begin by showing differences in outcomes for male versus female officers—specifically the differences in the percentages of male and female officers who are retained or promoted at each milestone. We then examine differences in the observed characteristics of male and female officers at each milestone, focusing on the milestones where we observe the largest differences in outcomes. Next, we show the results of the pooled regression model of the likelihood of achieving each milestone. By *pooled*, we mean including both male and female officers. As we show in Chapter Four, the Blinder-Oaxaca decomposition results will depend not only on differences in observed characteristics but also on the pooled regression estimates.

Differences in Mean Outcomes

Table 3.1 shows the differences in career progress for male and female officers. Specifically, it shows percentages of male officers and female officers reaching each milestone and the differences between percentages, not controlling for other factors that could be correlated with career progression. Thus, these are raw means.

The table shows large differences in the percentages of male and female officers in the O3 promotion window who are subsequently promoted to O3 and large differences in the retention of those promoted to O3. Specifically, we find a difference of 3.6 percentage points in the percentage of male versus female officers promoted to O3, condi-

Table 3.1
Officers Retained or Promoted in Phase 2 Analysis: Male Versus Female Officers

Milestone	Overall (%)	Male Officers (%)	Female Officers (%)	Difference (Male Minus Female)
Retained as O1	99.9	99.9	99.8	0.1
Promotion to O2	98.1	98.2	97.2	1.0
Retained as O2	99.5	99.6	99.2	0.4
Promotion to O3	92.7	93.3	89.6	3.6
Retained as O3	82.0	83.2	71.6	11.8
Promotion to O4	83.6	84.2	78.1	6.0
Retained as O4	92.3	92.4	91.3	1.2
Promotion to O5	85.4	85.6	83.5	2.1
Retained as O5	86.2	87.2	76.3	10.9
Promotion to O6	59.2	59.3	58.0	1.3

NOTE: The last column was calculated before the percentages were rounded. The largest gender differences are highlighted.

tional on O3 promotion eligibility. Given a promotion to O3, we find a difference of 11.8 percentage points in the retention of male versus female officers as an O3. Thus, fewer female officers are promoted to O3, and, of those promoted, fewer stay. Furthermore, of those who stay as an O3, considerably fewer female officers are subsequently promoted to O4, a difference of 6.0 percentage points. These results are consistent with those found in our first study. As in that study, we find that female officers have slower career progression in the company grades—primarily because of lower retention but also because of lower promotion.

Conditional on making the field grade of O4, we find that career progression between male and female officers is broadly similar, until O5, in terms of promotion and retention; we find that percentages of male officers who are retained as an O4 and who are promoted to O5 are slightly greater than the percentages of female officers. However, the percentage of male officers who are retained as an O5, conditional on promotion to that grade, is significantly higher, a difference of 10.9 percentage points. Thus, female officers are more likely to leave after reaching O5. Given that those who retire with 20 years of service are generally an O5, the O5 retention results suggest that female officers have a higher retirement rate than male officers, though it is important to note that the results in Table 3.1 show retention over all years of service that an individual is an O5 (up to the O6 promotion window). Among those who stay after reaching O5, the promotion rate to O6 is slightly less for female officers, 57.7 percent, compared with 58.8 percent for male officers.

In short, like the stage 1 analysis, we find that female officers experience different career progression than male officers—primarily because of lower retention and, to a lesser extent, lower promotion rates to O3 and O4. The purpose of the stage 2 analysis is to assess the extent to which these differences are attributable to differences in observed characteristics, such as family factors and occupational group. Before we make that assessment, we first examine the extent to which observed characteristics differ between male and female officers at each career milestone.

Mean Characteristics at Each Milestone

Our analysis considers observed job-related characteristics, including source of commission, occupational group, cumulative months deployed, and service branch, as well as individual characteristics, including race and ethnicity, education, marital status, and existence and age of dependents. For each career milestone, we computed the percentage of all officers at that milestone (or, in the case of months deployed, the mean number of months) with a given characteristic, the percentage of male officers, and then the percentage of female officers. The samples used to compute these means are the ones that are the basis for the regression analysis and the Blinder-Oaxaca decomposition.

Because we consider ten milestones and a large number of characteristics, the discussion here highlights a selected set of characteristics.

Appendix B presents a table of descriptive statistics (Table B.1) that gives all of the mean characteristics for each milestone for all officers, male officers, and female officers.

It is important to recognize that the tabulations are intended as background and to assist in the interpretation of the regression results in Chapter Four, where we consider to what extent observed gender differences in reaching key milestones are attributable to differences in observed characteristics at those milestones. The tabulations themselves are not intended to provide information on how the characteristics of an entry cohort of officers changed through each subsequent retention and promotion milestone. Differences in characteristics across career milestones, presented in Figures 3.1–3.4, are due to differences in retention and promotion across milestones *and* cohort effects. That is, the figures do not show differences in characteristics for a given cohort but for many cohorts. This is because our data draw from multiple cohorts and show the characteristics of those cohorts at each milestone over our data period, 2000–2010. Consequently, characteristics of those at the O4 to O6 milestones are those who entered prior to 2000, given that it usually takes at least ten years to achieve those milestones. As was shown in our earlier report (Asch, Miller, and Malchiodi, 2012), entry characteristics differ across cohort. Importantly, the representation of female officers was lower among earlier entry cohorts and less likely to be academy graduates and more likely to be in administrative occupations. For that reason, we must be careful not to interpret the results shown in Figures 3.1–3.4 as reflecting how characteristics change as members of cohort moves through their careers. Instead, we must interpret the results as the characteristics at each milestone, reflecting both changes as the cohort moves and differences in characteristics across cohorts.

Job-Related Characteristics

Figure 3.1 shows the percentage of officers with a given source of commission at each milestone. For example, the figure shows that, at entry, 24.9 percent of male officers are academy graduates, 36.0 percent of male officers have Reserve Officer Training Corps (ROTC) as their source of commission, and 29.2 percent have Officer Candidate School

Figure 3.1
Officers with a Given Source of Commission at Each Milestone

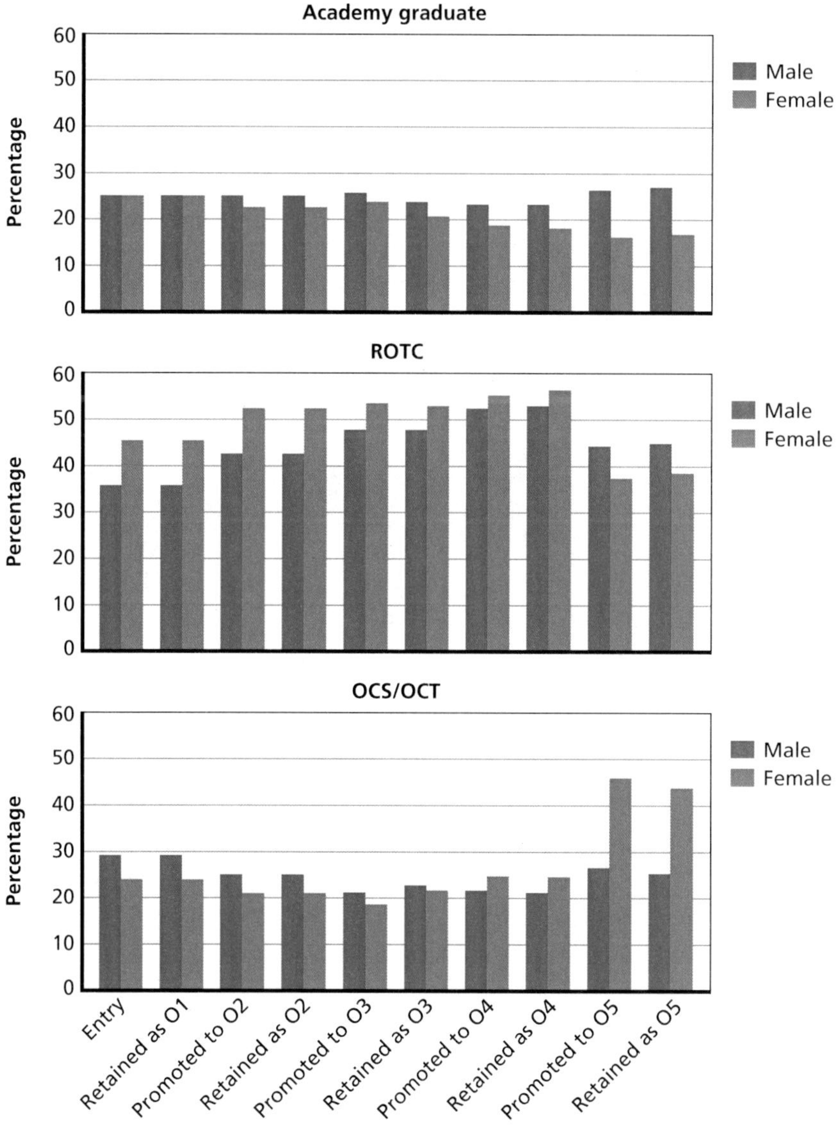

RAND *RR1288-3.1*

or Officer Candidate Training (OCS/OCT) as their source of commission.[1] The corresponding figures for female officers at entry are 24.9 percent, 45.6 percent, and 24.0 percent, respectively.

The top panel in the figure shows that while the percentages of male and female officers who are academy graduates are the same at entry, a smaller percentage of females are academy graduates at later career milestones, and the gap between the male and female percentage is larger at subsequent career milestones, especially at O5. For example, among those recently promoted to O4, 23.0 percent of male officers are academy graduates, compared with 18.5 percent of females, a difference of 4.5 percentage points. Among those recently promoted to O5, the difference is larger, 10.5 percentage points (26.4 percent for men minus 15.9 percent for women). Again, it is important to reiterate that the smaller percentage of female academy graduates at later career milestones could fully or partially reflect the lower representation of female academy graduates among less recent entry cohorts of officers.

The bottom two panels in Figure 3.1 show that the larger gap in the field grades between male and female officers in the representation of academy graduates, shown in the top panel, is not due to greater representation of ROTC graduates among female officers; the gap is due to greater representation of OCS/OCT graduates. In fact, we find that representation of ROTC graduates among female officers is greater than for males among those recently promoted to O4 (55.4 percent versus 52.3 percent) but is lower among those recently promoted to O5 (37.4 percent for females versus 44.5 percent for males). However, OCS/OCT representation among male and female officers is greater for those recently promoted to O4 and to O5 and dramatically larger among women recently promoted to O5. Thus, among women recently promoted to O5, nearly half—45.9 percent—are OCS/OCT graduates, while 37.4 percent are ROTC graduates and 15.9 percent are academy graduates.

These differences in source of commission between male and female officers among those recently promoted to O4 and to O5 are also

---

[1] The figures do not sum to 100 percent because some officers have other sources of commission.

generally observed among those who are retained as an O4 until the O5 promotion eligibility window and among those who are retained as an O5 until the O6 promotion eligibility window. For example, we find that 43.8 percent of women retained as an O5 are OCS/OCT graduates, just slightly less than the 45.9 percent of women who were recently promoted to O5.

Figure 3.2 shows the percentage of male and female officers at each career milestone who are in each DoD occupational group. The overall pattern is that tactical occupations are underrepresented among female officers, while the other groups, especially the administration DoD occupational group, are overrepresented.

In contrast, the representation of administrative occupations is greater at later career milestones for female officers, as seen in the second panel in Figure 3.2, again potentially reflecting the greater representation of this occupational group among earlier entry cohorts. In the administrative occupations, representation is generally stable across the career pipeline for men, so that the difference in representation of administrative occupations is larger later in the career, especially in the field grades. In short, in the field grades, female officers are far more likely to be administration than in tactical occupations, compared with O3, when female officers are more likely to be in tactical occupations.

In terms of how representation varies across subsequent career milestones, the top panel shows that, among men, the representation of tactical occupations is larger later in the career, especially in the company grades. For example, representation of tactical occupations among male officers is greater later in the career—28.0 percent at entry and 57.8 percent among those recently promoted to O3. In the field grades of O4 and O5, the representation among male officers stabilizes at around 63 percent. For female officers, representation of tactical occupations also increases during the company grades, albeit the level is lower than for male officers; unlike male officers, representation drops sharply at the O4 and O5 levels. For example, 27.8 percent of female officers who are retained as an O3 are in tactical occupations, but this figure drops to 16.5 percent among those recently promoted to O4 and drops further to 12.2 percent among those recently promoted

Figure 3.2
Officers in Each DoD Occupational Group at Each Milestone

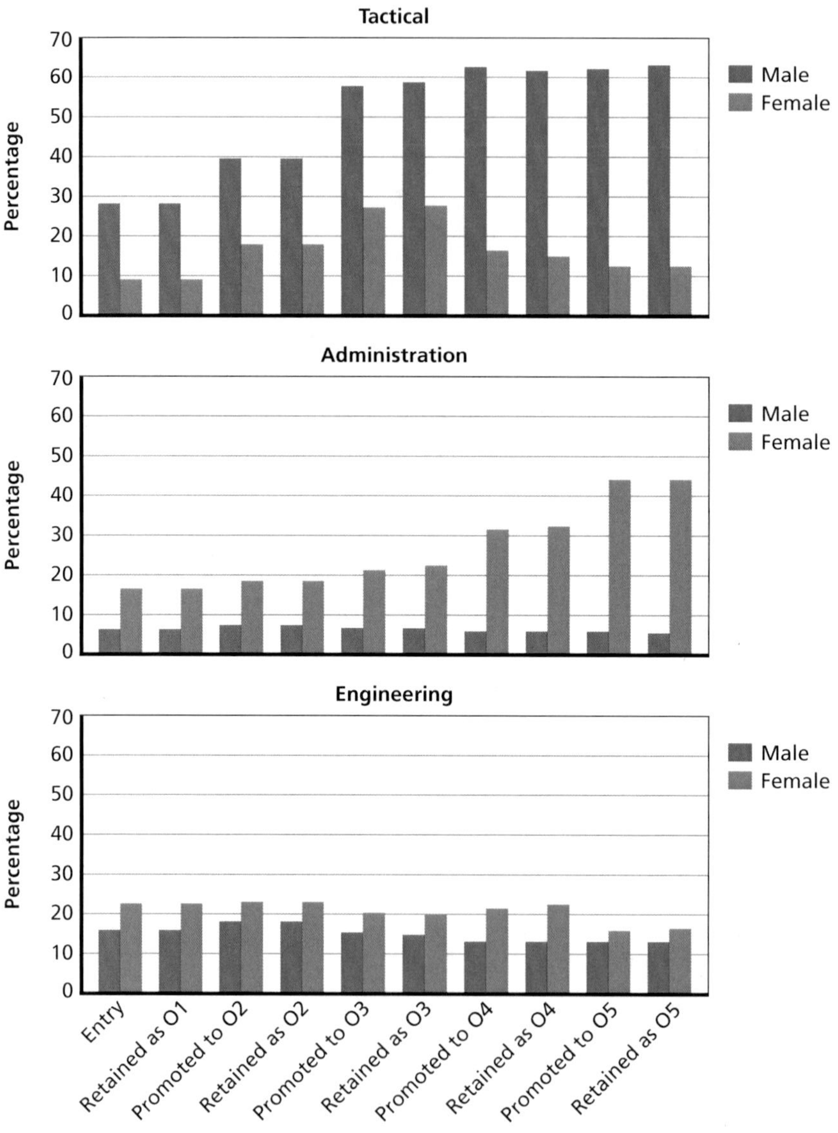

RAND *RR1288-3.2a*

Figure 3.2—Continued

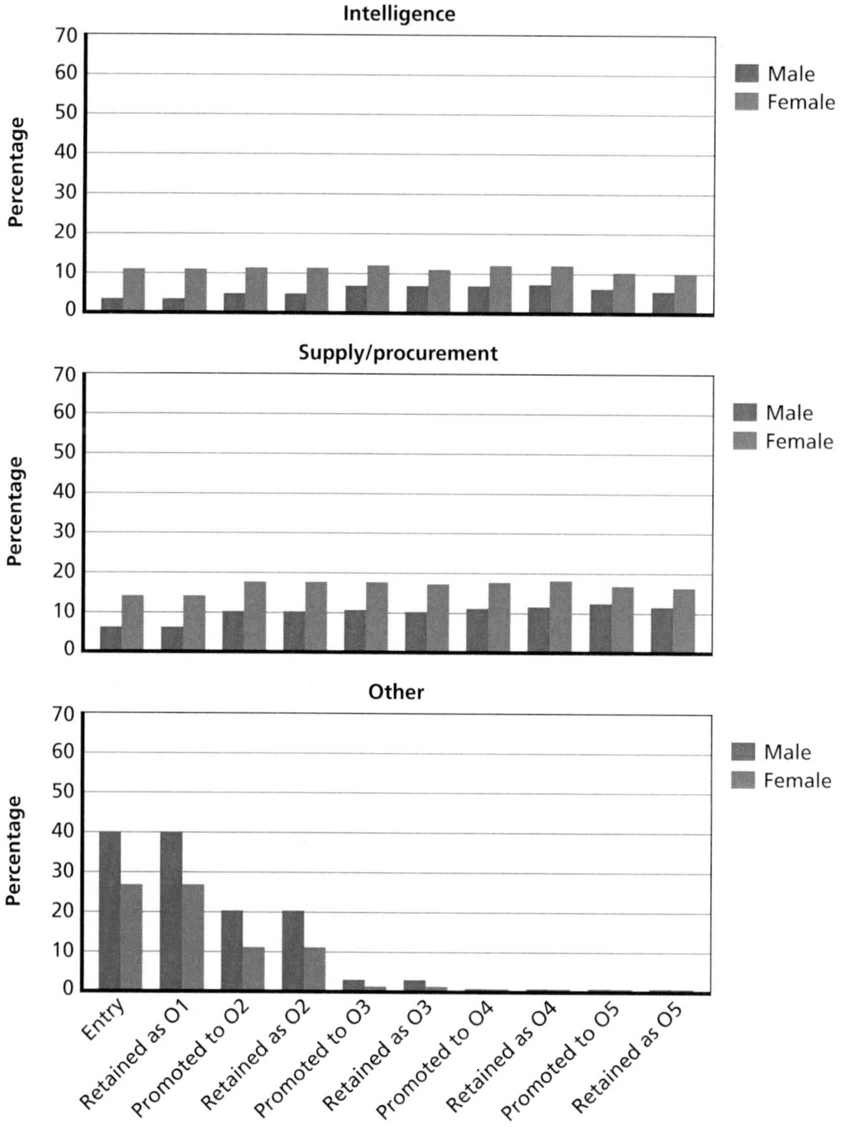

RAND *RR1288-3.2b*

to O5. Thus, the difference in the representation of tactical occupations between male and female officers is larger later in the career pipeline.

It should be recalled that the tabulations in Figures 3.1 and 3.2 are raw percentages and do not hold constant other characteristics that could affect representation of occupational groups. For example, the differences in source of commission observed in Figure 3.1 could affect the representation of occupational groups insofar as academy, OCS/OCT, and ROTC graduates are not equally represented in all occupations. That is, the lower representation of tactical specialties among women could reflect the lower representation of academy graduates among female officers beyond the entry point, because those in tactical occupations are more likely to be academy graduates.

Similarly, the differences we observe in representation among male and female company grade versus field grade officers could reflect differences in the entry cohorts used in our analysis. As noted, given that we only include observations from 2000 to 2010, those in more-junior grades are likely to have entered service more recently than those in more-senior grades. Thus, the differences across subsequent career milestones could reflect differences in the entry characteristics of recent versus older entry cohorts. For example, representation of academy graduates might have been greater among more-recent female entrants relative to later female entry cohorts and could account for the drop in the representation of academy graduates among the field grade relative to the company grades, as observed in Figure 3.1.

These observations highlight the importance of controlling for confounding factors, such as source of commission and entry cohort, in our regression analysis.

Table B.1 in Appendix B shows tabulations of other job-related characteristics that we consider in our analysis. As shown there, we find that the average cumulative months of deployment increase over the career, unsurprisingly, but that the average number of months is less for female than for male officers at each milestone. Similar to the pattern in Figure 3.2 for tactical specialties, the gap between male and female officers is greater in the more-senior field grades. We also consider the percentage of officers with prior enlisted service. This percentage is lower at higher grades for both male and female officers and is lower for female than male officers. That said, the difference in the percentage with prior service, among female and male officers, declines with grade; the difference is 1.1 percentage points (13.2 percent for men and 12.1 percent for women) among those retained as an O5, compared with 6.1 percentage points among those at entry (29.3 percent for men and 23.1 percent for women).

Individual-Related Characteristics

We next consider individual characteristics of officers, highlighting marital status and dependents. Again, our full tabulations are in Table B.1 in Appendix B.

Figure 3.3 shows the percentage of officers who are married at each career milestone. The percentage of officers who are married is greater among later career milestones for both male and female officers, but female officers are less likely to be married at each milestone. Furthermore, the gap between male and female officers is greater at later milestones. Specifically, at entry, 39.5 percent of female officers and 48.2 percent of male officers were married, a difference of 8.7 percentage points. Among those promoted to O3, given that they were retained

Figure 3.3
Officers Who Are Married at Each Milestone

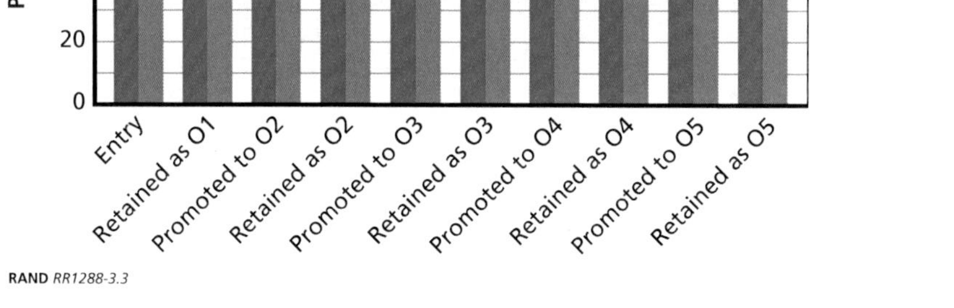

RAND *RR1288-3.3*

as an O2, the difference is 18.8 percentage points (62.9 percent for female officers versus 81.7 percent for male officers). The gap is even larger among those who were retained as an O3, given that there was first a promotion to O3, equal to 22.0 percentage points (63.7 percent versus 85.7 percent for female and male officers, respectively). Among the field grades of O4 and O5, the percentage married among female officers is roughly the same, around 68 percent, but is even higher for male officers. Thus, among those who were retained as an O5, almost all male officers were married, while only about two-thirds of female officers were. The two-thirds figure for female officers is higher than what was observed for earlier cohorts. For example, Hosek et al. (2001) found a marital rate for female officers at the later career milestones of around 55 percent.

We also consider the incidence of dual-military service among married personnel across subsequent career milestones. As shown in Figure 3.4, about 3 to 4 percent of male officers who were married were married to another service member. Female officers were far more likely to be married to another service member, and the incidence of such marriages varies across milestones. The percentage of married

Figure 3.4
Officers Who Have a Spouse in the Military, Among Those Married, at Each Milestone

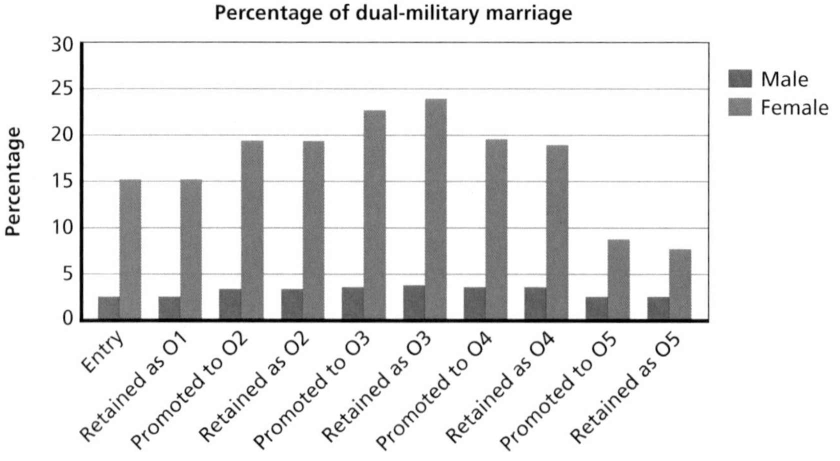

RAND *RR1288-3.4*

female officers married to another service member is greatest among O3s, both those recently promoted to O3 (22.7 percent of female officers) and those retained as an O3 (23.9 percent). This figure decreases at subsequent career milestones. Among married female officers recently promoted to O5, only 8.7 percent were married to another member, and among those retained as an O5, the figure is 7.8 percent.

The review of the literature pointed to the potential importance of children as a factor affecting female officers' retention and promotion outcomes. Figure 3.5 shows the percentage of male and female officers at each career milestone who have any dependents. Table B.1 shows the percentages with dependents in the age groups of birth to three, four to six, seven to 18, and 19 and older. The figure shows that the percentage of officers with dependents is higher at subsequent career milestones, though it flattens out for female officers for those at O4 and O5 milestones. That said, the percentage is lower among female officers, with the gap between male and female officers larger at higher grades. Thus, among those promoted to O5, given that they were retained as an O4, 56.1 percent of female officers had dependents, compared with

Figure 3.5
Officers with a Dependent, at Each Milestone

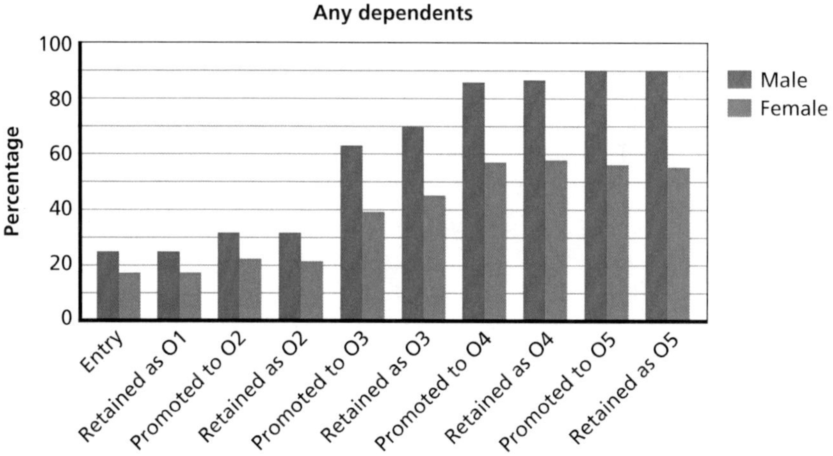

RAND *RR1288-3.5*

90.5 percent of male officers. This is compared with 17.3 percent of female officers at entry versus 25.2 percent for men.

The tabulations in Table B.1 show that the ages of dependents vary across career milestones for both male and female officers.[2] The percentage of both male and female officers with very young dependents, ages birth to three, is greatest among O3s, though the percentage is higher for men than for women even among O3s. Similarly, the percentage of officers with dependents ages four to six peaks for both male and female officers at O4; the percentage of officers with dependents ages seven to 18 peaks at O5 for both male and female officers, though in each case the percentage is lower for female officers. The results for the incidence of adult dependents over age 18 differ slightly, however. Here, we find that the incidence peaks at O5, as in the case of the incidence of dependents ages seven to 18, but unlike this younger category, the incidence is slightly higher for female officers than for male officers for grades O1 to O4. That is, a slightly higher percentage of female officers in grades O1 to O4 had adult dependents, but among those at O5, a much higher percentage of male officers than female officers had adult dependents. Specifically, among those retained as an O4, 12.6 percent of female officers had adult dependents, compared with 11.1 percent of male officers, but among those promoted to O5, these percentages are 17.0 percent and 29.2 percent, respectively. It is unclear what drives the differing pattern for the incidence of adult children versus younger children. One possibility is that the pattern for younger children reflects childbearing at younger ages among officers and the aging of those children as officers progress in their careers, while the pattern for adult children could reflect both the aging of children and the incidence of divorce and subsequent remarriage with new spouses with adult children.

In addition to marriage and dependents status, we also consider education, including the percentage of officers with only a bachelor's degree versus an advanced degree, and race and ethnicity. In the case of education, the percentage of officers with an advanced degree is dramatically higher in the field grades relative to the company grades, and

---

[2] In our analysis, dependents do not include spouses.

the percentage among those in the field grades is higher for female officers relative to male officers. Thus, the incidence of advanced degrees is higher among female officers.

Pooled Regression Results

To perform the Blinder-Oaxaca decomposition, we must estimate the vector of estimated coefficients from the pooled regression for each milestone, or β^*, in the regression equation $Y = X\beta^* + v$ (see Appendix A). Specifically, we estimate linear probability models of achieving each milestone using ordinary least squares.[3] The variable Y is a zero-one indicator variable that equals 1 if the officer achieves the milestone, conditional on achieving the previous one, and zero otherwise. It is a pooled regression because we pool together data for male and female officers. The pooled regression results for each milestone are shown in Table B.2 in Appendix B. Note that the pooled regressions are estimated with the constraint that the indicator variables sum to one, to address the problem of the choice of omitted group in defining categorical variables for the Blinder-Oaxaca decomposition. This constraint results in transformed or normalized estimates on the categorical variables, and the interpretation of the coefficient estimate is the deviation from the grand mean of the coefficients for all categories. Thus, the results in Table B.2 show the estimates with the normalized

---

[3] Many have argued for the use of limited dependent variable models for binary choice, such as the probit and logit, instead of the linear probability model. These arguments are typically made on two grounds: (1) ordinary least squares imposes heteroscedasticity in the case of a binary outcome, and (2) linear probability estimates are not constrained to the unit interval. We can and do easily address the first argument by using heteroscedasticity-robust standard errors. While the second argument is more concerning and a natural and unattractive feature of the linear probability model, it is well-established that any bias stemming from this issue dissipates and approaches zero as the share of predicted probabilities falling outside the unit interval approaches zero (Horace and Oaxaca, 2006). Thus, the linear probability model is generally considered appropriate in cases where the mean outcome is not close to zero or one. While we do apply our model to all milestones, we are primarily interested in those milestones that have the lowest rates of attainment, and hence are furthest away from one. More specifically, we limit our discussion to outcomes with attainment rates significantly below one: retention at O3, promotion to O4, retention at O5, and promotion to O6.

coefficient estimates. As explained in Chapter Two, these regression estimates together with the mean differences in characteristics are the inputs to the Blinder-Oaxaca decompositions (shown in the next chapter). Here, we present a brief overview of some of the key results.

As shown earlier this chapter, female officers are less likely to be academy graduates and more likely to be ROTC graduates, except in the later career, when they are more likely to be OCS/OCT graduates. In the pooled regressions, we find that being an academy graduate is positively related to reaching a given milestone, except in the midcareer; this is seen in the likelihood of being retained as an O3 and being promoted to O4. On the other hand, being an OCS/OCT graduate is negatively associated with the probability of reaching later milestones, including promotion to O5 and retention as an O5. As we show in Chapter Four, some of the explainable differences in reaching these later career milestones are attributable to commissioning source.

We also showed that female officers are far less likely than male officers to be in tactical occupations and far more likely to be in administrative occupations, especially later in the career. Officers in administrative occupations have a lower probability of reaching early career milestones, through promotion to O3, but we generally find no statistically different probability of reaching later career milestones, relative to the overall average effect of all occupation groups. The exception is the finding that the probability of being retained as an O5, given that it is preceded by O5 promotion, is positively associated with being in an administrative occupation. We find that generally being in a tactical occupation is positively associated with the probability of achieving each milestone except O4, where the likelihood of being promoted to O4 is negatively related to being in a tactical occupation and is negatively related to the probability of being retained as an O4, conditional on promotion until that point. We find that occupational group is among the group of characteristics that tend to contribute the most to the observable differences between male and female officers in the probability of reaching various career milestones, as shown in Chapter Four.

The cumulative months of deployment is another characteristic we consider. Female officers have fewer months, especially at later career milestones, and we find that having more months is positively related to achieving retention and promotion milestones, with the exception of retention as an O3 and promotion to O5. For these milestones, the probability of attaining these milestones are negatively related to cumulative months of deployment.

We find that female officers are more likely than male officers to be in the Army and Air Force during early career milestones but less likely to be in the Army during later career milestones and about equally likely to be in the Air Force later in the career. Thus, the distribution of female to male officers across service branch varies over the career. These percentages are shown in Table B.1. In the pooled regressions, we find that being in the Air Force is positively associated or has no statistically significant association with the likelihood of reaching each milestone. In contrast, being in the Navy often has a negative association. That is, relative to the overall average effect across the services, naval officers generally have a lower likelihood of achieving each milestone. In the Marine Corps and Army, the relationship is mixed—sometimes negative, sometimes positive.

In terms of individual characteristics, female officers were less likely to be married at each milestone and more likely to be a dual-military spouse. In the pooled regressions, we find that married personnel were more likely to be promoted to each milestone and less likely to be retained, relative to single personnel. As we show in Chapter Four, the difference in marital status between male and female officers is a key contributor to the observable differences in their career pipelines. Similarly, female officers were less likely to have dependents, and we find that, in general, having dependents of any age was positively related to the probability of reaching a given milestone. The exception is retention as an O5, where the likelihood of being retained as an O5 was negatively related to having dependents younger than 18 but positively related to having dependents 19 or older.

CHAPTER FOUR

Decomposition of Gender Differences in the Likelihood of Achieving Career Milestones

To what extent do differences in characteristics explain gender differences in career progression, and which specific characteristics have the most important role in contributing to differences in observables versus associations? We address these questions in this chapter by showing results of the Blinder-Oaxaca decomposition approach, described in Chapter Two and Appendix A. The characteristics we consider are marital status, joint-duty status, number of children, ages of children, entry cohort year, service branch, source of commission, prior enlisted service, occupational group, cumulative months of deployment, education, race, and ethnicity.

We begin the discussion by first showing the decomposition of gender differences in the probability of reaching each milestone into the part that is attributable to differences in observables versus differences in associations. As discussed in Chapter Two, differences in *observables* refer to the part of the gender difference in outcomes attributable to differences in observed characteristics, while differences in *associations* refer to the part attributable to gender differences in the association of the characteristic and the outcome. The second part shows the differences in the outcomes when male and female officers have the same observed characteristics. This discussion provides information on the relative importance of observed differences versus differences in associations in contributing to the overall gender gap in the probability of reaching each milestone.

We then further decompose the observable and association components into the contributions of each individual characteristic for selected milestones. The decomposition of the gender difference in the probability of reaching each milestone is shown in Appendix B. In the main text, we show the results graphically.[1] The chapter concludes with a discussion of the overall contribution of specific factors, focusing on the probability of reaching milestones with the largest gender gap.

Decomposition into the Observable and Association Components

Table 3.1 in Chapter Three showed the probability of reaching each milestone for male and female officers and the difference in these probabilities. We used the Blinder-Oaxaca methodology to decompose the differences shown the last column of Table 3.1 into an observable and association components.

To illustrate the method, Table 4.1 shows an example of the decomposition, using retention as an O3. The left three columns of the table replicate information from Table 3.1. Male officers have 83.2-percent likelihood of being retained as an O3 until the O4 promotion window, conditional on promotion to O3, while female officers have 71.6-percent likelihood, for a difference of 11.8 percentage points. Using the Blinder-Oaxaca decomposition method, the portion of the 11.8-percentage-point difference that is due to differences in observables is 7.9 percentage points. That is, differences in observed characteristics account for 7.9 percentage points out of the 11.8-percent total, while 3.9 percentage points are attributable to the unexplained part. Of the 7.9 percentage points, 2.9 percentage points, or about one-

---

[1] Given that different services have adopted different policies and programs to help address gender differences in officer career progression, we checked to see whether the patterns in contributors to explained and unexplained gender differences in officer career progression varied by service. We did this by running our models specifically for the Navy. The results for Navy officers are qualitatively similar to those for the military as a whole, which suggests that patterns are likely similar across services. We thus present results for the military as a whole, but results for each service are available upon request.

Table 4.1
Example of the Decomposition of the Difference of Officers Retained as an O3

Milestone	Male Officers (%)	Female Officers (%)	Difference (Male Minus Female)	Part Due to Differences in Observables (%)			Part Due to Differences in Associations (%)
				Overall	Part Due to Having Any Dependents	Part Due to Having Any Dependent Ages 0–3	
Retained as O3	83.2	71.6	11.8	7.9	2.9	1.0	3.9

third of the observable difference, is attributable to gender differences in the percentage of officers with dependents. As shown in Chapter Three, female officers were less likely to have dependents. Furthermore, 1 percentage point is due to differences in the percentage with young dependents (birth to three). Thus, of the part of the gender difference that can be explained (7.9 percentage points), about one-third (2.9 percentage points) is due to differences in the presence of children, and about one-third of that (1 percentage points) is due to differences in the presence of babies and toddlers. Female officers were less likely to have young children, and those with young children were more likely to have been retained as an O3.

In our example, differences in the presence of young children for male and female officers contributed positively to the observable gender difference in the likelihood of being retained as an O3. Because the difference in the O3 retention milestone is positive—men were more likely to be retained as an O3—and because small children contributed positively to that difference, the difference in this characteristic was therefore disadvantageous to female officers. Put differently, the fact that female officers were less likely to have small children than their male counterparts helps explain why women are more likely to leave before being promoted to O4.

As a reminder, our results should be considered descriptive and not causal and can reflect selection factors and endogeneity of the marriage and childbearing decisions. For example, it is possible that female officers are less likely to have children at the O3 milestone, because those with children are less likely to be promoted or retained at earlier milestones (possibly leaving the military at an earlier milestone to have children). We consider this issue later in the chapter.

Figure 4.1 shows the decomposition of the gender difference in the probability of reaching specific milestones into the two main components: differences in observables characteristics and differences in associations. The largest gender differences in the probability of reaching a given milestone occur at promotion to O3, retention as an O3, promotion to O4, and retention as an O5. The discussion in this section focuses on these, as well as the final milestone, promotion to O6, because of the relatively large size of the two components.

Figure 4.1
Probability of Achieving Key Career Milestones: Decomposition of the Gender Differences Attributable to Observable and Association Components

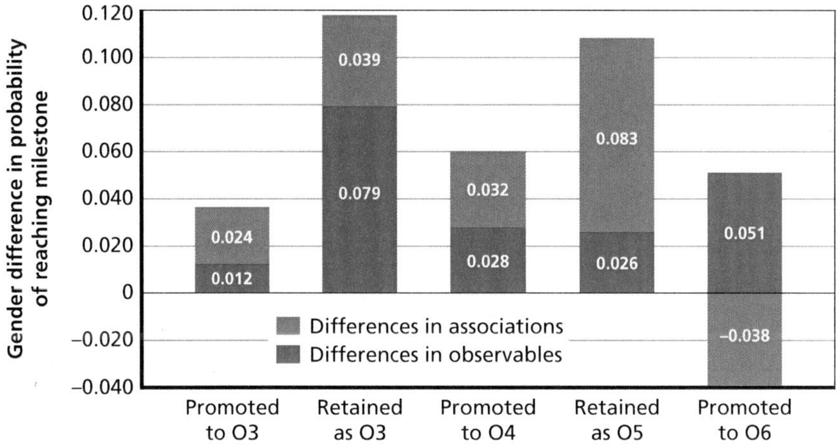

NOTE: A positive difference in the figure is disadvantageous for female officers because it contributed to a larger gender difference in career progression.
RAND *RR1288-4.1*

We find that differences in observed characteristics account for only part of the gender differences in the officer career pipeline. In the case of retention as an O3, the mean gender difference is 11.8 percentage points in the probability of achieving this milestone, with about two-thirds of this difference attributable to observed factors, or 7.9 percentage points. In the case of promotion to O4, differences in observed characteristics explain only about half of the mean gender difference in the probability of reaching this milestone (0.028 of 0.06). In the case of gender differences in the probability of being retained as an O5, observed factors account for about one-quarter of the overall difference (0.026 out of 0.109), implying that differences in associations account for about three-quarters of the observed difference. Thus, observed factors play a less prominent role at the later retention milestone.

The mean gender difference in the probability of being promoted to O6 is only about 1 percentage point (see Table 3.1). Figure 4.1 shows that while observable factors contributed positively to the gender difference, differences in associations contributed negatively by an almost offsetting amount. This suggests that female officers at the O6 promotion point had observed characteristics that worked against them relative to male officers, but differences in the association between characteristics and promotion worked for them.

Further Decomposing the Observable Components into Contributions Because of Differences in Specific Characteristics

Our analysis is designed to answer not only how much of the gender difference is due to differences in observed characteristics but which specific characteristics contributed the most to this difference. We further decompose the observable differences shown in Figure 4.1 into the contributions of differences in individual observed characteristics, as explained in Appendix A. We show the results graphically, focusing on the key milestones shown in Figure 4.1, with the full results provided in Appendix B.

Three issues should be kept in mind regarding this analysis. First, to reduce clutter in the graphics, we show the combined contribution of categorical variables. The combined contribution is the sum of the contributions of the subcategories. For example, we show the contribution of occupational group, equal to the sum of the contribution of being in each DoD occupational group (tactical, intelligence, engineering, supply and procurement, administration, and other). The categorical variables for which we show the combined effects are source of commission, entry year, occupational group, race and ethnicity, and service branch. We also combined marital status and any dependents to create a family-status variable.

Second, as discussed in the context of Table 4.1, some characteristics contributed positively, while others contributed negatively to the gender difference in the probability of achieving a given milestone. A positive contribution means that the gender difference in the characteristic, together with the estimated effect of the characteristic, increased the observable component of the gender difference in the probability of achieving the milestone. That is, a positive contribution means that the difference in the characteristic is disadvantageous to female officers. A negative contribution means that the gender difference in the characteristic is advantageous to female officers because it reduces the difference between male and female officers in the probability of reaching a given milestone, attributable to observed factors.

Finally, we use the same scale across milestones to show the results of the detailed decompositions. Using the same scale facilitates comparisons of the relative importance of specific characteristics across milestones. One result of the use of a common scale is that some of the decompositions will appear to be rather "squashed."

We first summarize the overall findings in Table 4.2, before showing specific results. The table shows the main contributors to the differences in the probability of reaching the selected milestones, attributable to observed differences in terms of magnitude and in terms of being statistically different from zero.

The table shows that both individual characteristics and job characteristics contributed to differences in observed characteristics. In terms of individual characteristics, family-related factors, specifically marital status and presence of children, are major contributors to the positive gender gap in career milestones attributable to observed characteristics. In particular, we find that female officers were less likely to be married and less likely to have children, whereas being married and having children were generally positively related to career progression. Thus, family status contributed positively to the observed component

Table 4.2
The Main Contributors to Gender Differences in Reaching Selected Milestones, Observed Characteristics

Key Milestone	Difference in Gender Gap Attributable to Differences in Observed Characteristics
Retained as O3	Family status, occupation, entry year
Promoted to O4	Family status, occupation, entry year
Retained as O5	Occupation
Promoted to O6	Family status, deployment experience

of the gender gap. These results are consistent with research findings on the work outcomes of civilian women with more education. Thus, as with their civilian counterparts, better-educated women in the military postponed marriage and childbirth. Yet, also like their civilian counterparts, married men had better career outcomes (A. Miller, 2011; Buckles, 2008; Lundberg and Rose, 2000; Waite, 1995).

We cannot infer from these results that family status causes gender differences in career progression. In fact, the results might reflect reverse causality. That is, female officers might be less likely to be married and be retained because they leave so that they can marry. Because those who leave the service to get married appear as single in our data, it can appear that retention is higher among those who marry. If this is the case, we should expect to find that single women were more likely to leave than married women. We find no evidence to support this argument. The estimated coefficient for marriage is not statistically significant in the O1, O2, or O5 retention regressions for female officers. It is statistically significant but the wrong sign in the O3 and O4 regressions for female officers.[2] We find that single female officers were more likely to stay, not leave, as an O3 and as an O4. Thus, while we find that the lower marriage rate among female officers contributed to the positive gender gap in O3 retention, we do not find evidence to suggest that the lower marriage rate occurred at the O3 level because single women leave as an O3.

A similar argument could also be made for young children, from birth to age three. Female officers might have left service to bear children, with the results appearing as if having young children improved retention relative to those without young children. In this case, we do find some evidence consistent with this argument at the O3 retention milestone. In the O3 retention regression for female officers (see Appendix B), we find that female officers with dependents ages zero to three had higher O3 retention relative to those without young dependents. Thus, it is possible that those without young dependents leave as an O3 to have additional young children. However, this result is also consistent with other arguments, including that military service pro-

2 The tables in Appendix B omit the results for O1, O2, and O4 retention.

vides a positive incentive to have young children, in the form of access to health care, thereby inducing greater retention. We find that having older children—ages four to six, ages seven to 18, and over 18—was also positively related to O3 retention, suggesting that children in general, not specifically having new children or young children, are positively associated with O3 retention. This lends some support to the argument that incentive argument for the relationship between children and O3 retention. That said, because we do not have a behavioral model of marriage formation or childbearing decisions, we are unable to distinguish between these arguments.

It is interesting to note that, unlike the O3 retention result, we find that having young children (birth to three) is negatively associated with retention for female officers at the O4 and, especially, O5 retention milestones. These results run counter to the argument that female officers without young children leave to have them. However, the negative coefficient estimate on the variable for children ages zero to three for female officers could suggest that higher-ranked military positions are not conducive to raising small children, and so these women are more likely to leave. It could also suggest that female officers with young children are more responsive to retirement benefits and so are more likely to leave, since officers are typically at the O5 level when they become eligible for these benefits. We note that the coefficient estimates for the variable for children ages zero to three are also negative for male officers, so these arguments could also be relevant to male officers. Again, we are unable to distinguish between the various explanations for our results.

In terms of job-related characteristics, differences in occupational groups are a major contributor—usually a positive contributor, though in the case of promotion to O4, a negative contributor—to differences in the gender gap in the probability of reaching career milestones attributable to observed characteristics. Related, since the amount of deployment will depend on occupation, we find that cumulative months of deployment were also a contributor, especially later in the career (promotion to O6); this characteristic is usually disadvantageous to female officers when it is a main contributor.

Results for Specific Milestones

We next show specific results for each milestone, first considering retention as an O3 to the O4 promotion window, conditional on promotion to O3. Figure 4.2 shows the decomposition of the 7.9-percentage-point difference in O3 retention between male and female officers that is due to observed characteristics. We find that most of this difference is attributable to family factors and occupational groups. Together, these two factors account for 85 percent of the difference in observables in the probability of retention as an O3, with 60 percent attributable to the lower incidence among female officers of marriage and the lower likelihood of having any dependents. In the case of dependents, 2.9 percentage points of the overall 7.9-percentage-point difference is attributable to differences in presence of children.

With respect to occupation, an additional 25 percent of the gender difference attributable to observed characteristics in the probability of being retained as an O3 is due to differences in the occupational distribution of male and female officers. Finally, differences in

Figure 4.2
Probability of Retention as an O3: Decomposition of the Part of the Gender Difference Attributable to Observed Characteristics

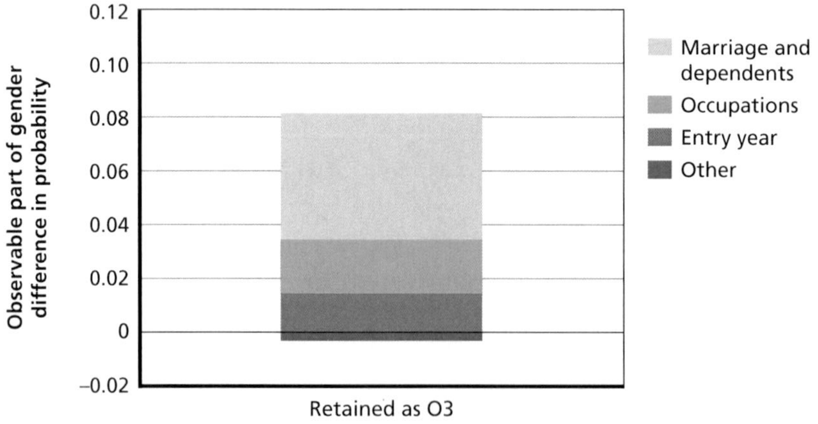

NOTE: The "other" category includes the contributions of prior service, race, ethnicity, commissioning source, service branch, education, and cumulative deployment.
RAND *RR1288-4.2*

entry year also contributed positively to the observable component in O3 retention. Female officers at the O3 retention decision point were more likely than male officers to have entered recently, and those who entered more recently had a lower probability of being retained. Thus, the contribution of entry year was disadvantageous to female officers at the career milestone of O3 retention.

As mentioned earlier, the graphs show the combined effects of categorical variables. But given the relative importance of dependents in the detailed decomposition in Figure 4.2, we further decomposed the "dependents" contribution into the contribution associated with specific age groups (Figure 4.3). The birth-to-three, four-to-six, and seven-to-18 age groups contributed about equally to the total, roughly one-third each. We also further decomposed the occupation category into the contribution of specific occupational groups (Figure 4.4). Similarly, after further decomposing the role of occupational group, we find that of the 1.9 percentage points attributable to occupations in Figure 4.2, the majority (1.2 percentage points) is due to the role of tactical occupations. Thus, the substantially lower probability of find-

Figure 4.3
Probability of Retention as an O3: Decomposition of the "Any Dependents" Part of the Gender Difference Attributable to Observed Characteristics

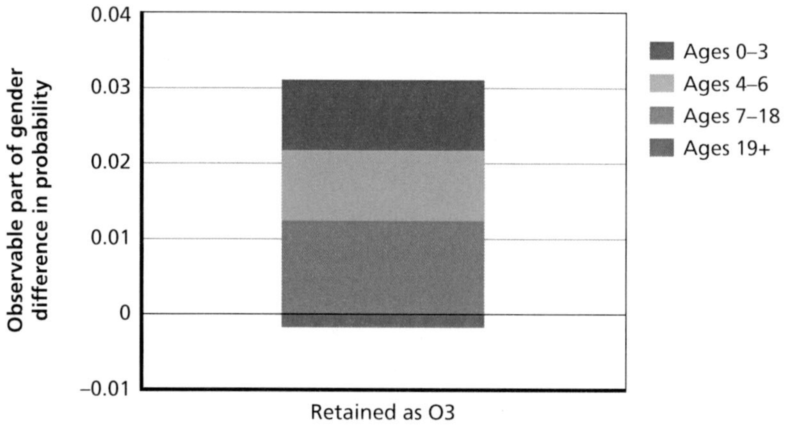

RAND *RR1288-4.3*

Figure 4.4
Probability of Retention as an O3: Decomposition of the "Occupations" Part of the Gender Difference Attributable to Observed Characteristics

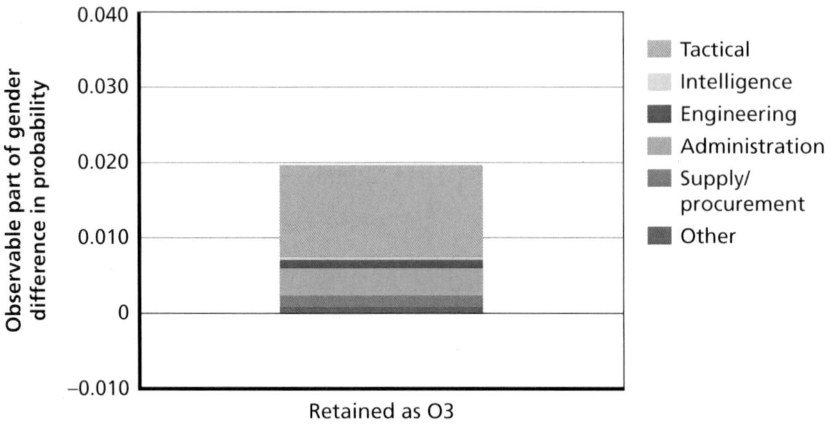

NOTE: The "other" category refers to all other occupational groups in the data, including science/professionals and those classified as nonoccupational.
RAND *RR1288-4.4*

ing female officers in tactical occupations and the positive relationship between being in a tactical occupation and retention as an O3 contributed importantly to the lower probability of retention as an O3 for female relative to male officers.

Figure 4.5 shows the detailed decomposition for the next milestone, probability of promotion to O4, conditional on retention as an O3. Here, entry year played a prominent role in contributing to the component attributable to differences in observed characteristics. Female officers had a higher representation in more-recent entry years, but these entry years are associated with lower promotion to O4, conditional on retention as an O3. Family status—specifically, differences in marital rate and presence of dependents between male and female officers—also contributed positively to the gender gap in the probability of an O4 promotion. However, gender differences in occupation contributed negatively and were advantageous to female officers in terms of promotion to O4.

Figure 4.5
Probability of Promotion to O4: Decomposition of the Part of the Gender Difference Attributable to Observed Characteristics

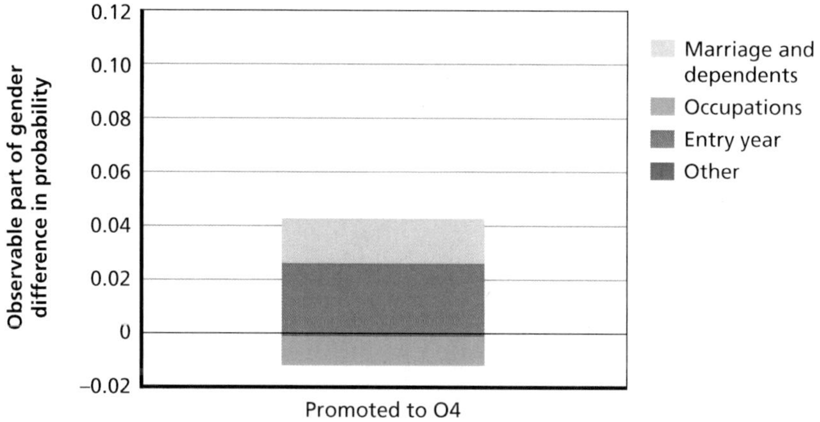

NOTE: The "other" category includes the contributions of prior service, race, ethnicity, commissioning source, service branch, education, and cumulative deployment.
RAND *RR1288-4.5*

The result for occupation (Figure 4.5) is surprising, given that occupation was a positive contributor to the gender gaps we observed, but here it was a negative contributor. Female officers eligible for an O4 promotion were less likely to be in tactical occupations, conditional on retention to the O4 promotion window, but unlike other milestones, we find that being in a tactical occupation was negatively associated with O4 promotion. This finding could be due to selectivity factors related to occupation that affect eligibility for O4 promotion. For example, those in administrative occupations were less likely to stay as an O3, but those who stayed might have more strongly preferred the military or had better O4 promotion prospects and so experienced greater probability of promotion.

Figure 4.6 shows the decomposition of the gender differences that is attributable to observed characteristics in the probability of being retained as an O5. The observable component is relatively small (as shown in Figure 4.1), with occupational differences being the main contributor.

Figure 4.6
Probability of Retention as an O5: Decomposition of the Part of the Gender Difference Attributable to Observed Characteristics

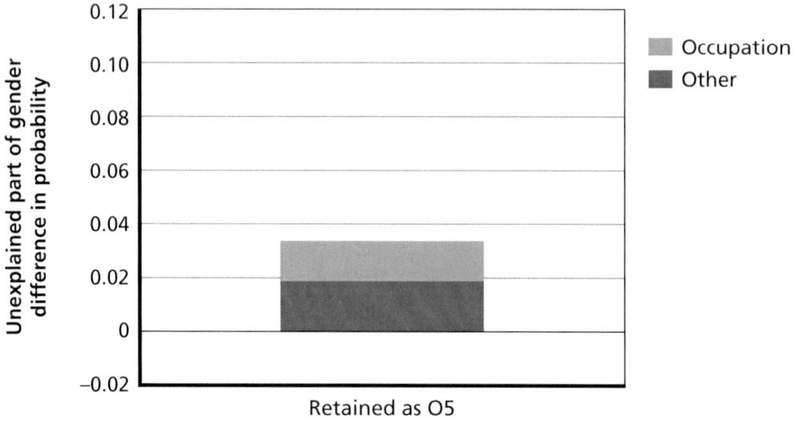

NOTE: The "other" category includes the contributions of prior service, entry year, dependents, marital status, race, ethnicity, commissioning source, service branch, education, and cumulative deployment.
RAND *RR1288-4.6*

The last milestone we consider is promotion to O6. As shown in Figure 4.1, most of the gender difference attributable to observed characteristics in the probability of reaching this milestone was offset by the portion attributable to the difference in associations. The latter portion is 5.1 percentage points. The two major contributors to the portion attributable to differences in observed characteristics are marital status and number of dependents and cumulative months of deployment, as shown in Figure 4.7. Female officers eligible for O6 promotion had less deployment experience and were less likely to be married. Both of the factors contributed the observable portion of the gender difference. For example, the contribution of cumulative deployment experience is 3.9 percentage points.

Decomposing the Contribution of Being a Dual-Military Spouse

Past studies have suggested that female officers married to other military members might face unique issues, compared with officers married to civilians, such as maintaining a joint domicile or ensuring adequate

Figure 4.7
Probability of Promotion to O6: Decomposition of the Part of the Gender Difference Attributable to Observed Characteristics

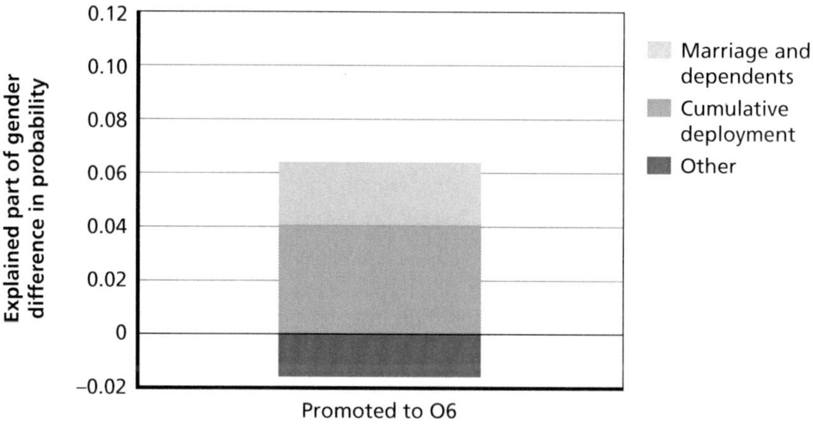

NOTE: The "other" category includes the contributions of dependents, prior service, occupations, commissioning source, race, ethnicity, service branch, entry year, and education.

RAND *RR1288-4.7*

childcare during deployments. Consequently, being a dual-military spouse might be a contributing factor to gender differences in the officer career pipeline. As shown in Figure 3.4, female officers were more likely to be married to a fellow service member over the career pipeline than male officers.

We examine the role of being a dual-military spouse, in the observable part of differences in career progression between male and female married officers, by further decomposing the role of marital status into the portion attributable to being married to a civilian versus another service member. Figure 4.8 shows how much of the gender difference in the probability of reaching each career milestone is attributable to differences in observed characteristics (similar to Figure 4.1) and how much of the observed differences are, in turn, attributable to being a dual-military spouse.

We find that being a dual-military spouse contributed little to the portion of the gender differences in the officer career pipeline that is attributable to differences in observed characteristics. In fact, except for

Figure 4.8
Probability of Achieving Key Career Milestones: Decomposition of the Gender Difference in the Observable Component and Part Attributable to Dual-Military Status

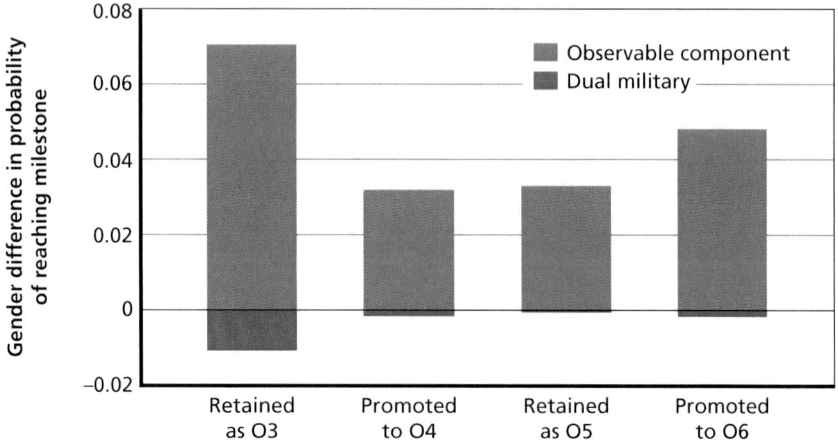

RAND *RR1288-4.8*

retention as an O3, dual-military status had no role in contributing to the observable portion of the gender gap for married officers. And, in the case of retention as an O3, the contribution of being a dual-military spouse is negative. In other words, being a dual-military spouse is advantageous to female officers for this career milestone.

This finding is surprising given the level of attention that issues and challenges facing dual-military couples have received from the research and policy communities (L. Miller et al., 2011; Smith, 2010; Moini, Zellman, and Gates, 2006; Steinberg, Harris, and Scarville, 1993; Teplitzky, 1988). Our study does not imply that dual-military couples do not face challenges with such issues as colocation and finding adequate and dependable childcare, as past research has demonstrated. However, it does suggest that such factors might not translate to material differences in the career progression of dual-military spouses. On the one hand, it could be that civilian spouses of officers tend to have high-stress professional careers that could also contribute to the lower retention rates of their officer spouses. On the other hand, it could be that dual-military spouses exhibit higher attachment to their military careers, which might counterbalance the negative influence of those challenges. Finally, it could be that the programs and policies that the services have developed to address issues for dual-military spouse, such as programs to prioritize the colocation of spouse duty, have helped to ameliorate those issues.

Further Decomposing the Association Components into Contributions of Specific Characteristics

We further decomposed the gender differences attributable to differences in the association of characteristics, shown in Figure 4.1, into the contributions of specific characteristics. That is, we held a given characteristic constant for male and female officers and asked to what extent gender differences in the association between that characteristic and the outcome contributed to the association component. Thus, this analysis provides information on the extent to which a given characteristic translated differently for male and female officers in terms of reaching a specific milestone, given that they were both eligible for that milestone. The association component is related to differences in the regression-coefficient estimates for female and male officers at each milestone, but those differences were normalized relative to the pooled regression-coefficient estimates and further normalized to account for the role of omitted categorical variables. Thus, while the differences in coefficient estimates for male and female officers are what we captured with the association component, the actual computation is more complicated (as shown in Appendix A), as is the interpretation of the component. We show the results graphically, focusing on the key milestones, shown in Figure 4.1, with the full results shown in Appendix B.

We first summarize the overall findings in Table 4.3, expanding on Table 4.2.

The main finding is that multiple factors contributed to the gender difference in terms of the probability of reaching the selected milestones attributable to differences in the association between outcomes and characteristics, with no single factor being the dominate contributor. That said, family status is a statistically significant contributing

Table 4.3
Summary of the Main Contributors to Gender Differences in Reaching Selected Milestones

Key Milestone	Difference in Gender Gap Attributable to Differences in Observed Characteristics	Difference in Gender Gap Attributable to Differences in the Association Between Characteristic and Outcome
Retained as O3	Family status, occupation, entry year	Family status, occupation, deployment experience, prior service, entry year
Promoted to O4	Family status, occupation, entry year	Family status, race/ethnicity
Retained as O5	Occupation	Prior service, entry year
Promoted to O6	Family status, deployment experience	Family status, deployment experience, race/ethnicity

factor to association differences in all of the milestones we considered, except retention as an O5. As we will discuss, few factors are statistically significant in the case of retention as an O5.

We find that male and female officers with the same family status—marital status and age and presence of dependents—had different likelihoods of being retained at O3 (conditional on being promoted to O3), being promoted to O4 (conditional on being retained as an O3), and being promoted to O6 (conditional on being retained as an O5). For example, we find that having dependents ages seven to 18 was positively associated with O6 promotion for male officers but negatively associated with O6 promotion for female officers. (It is important to recall that the different likelihoods for male and female officers with the same characteristics could be attributable to differential self-selection and gender differences in the causal effects of family on retention and promotion.)

More specifically, we find that among married officers, men had higher retention and promotion rates than women, holding other observed characteristics constant, contributing to the positive gender gap in retention and promotion rates between male and female officers. In only one case is marital status statistically significant for female officers—at the O3 retention milestone—and the estimated coefficient is negative, indicating that married female officers were more likely to leave. In contrast, marital status is positive and statistically significant for male officers at every milestone.

The role of very young children (birth to three), varies at the different milestones. We find that having very young dependents was positively associated with O3 retention for both male and female officers, but the relationship is stronger for males. That is, those with young children have higher O3 retention, but more so for men. The weaker relationship for women contributed positively (though not statistically significantly) to the O3 retention gap between male and female officers. In the case of O5 retention, having very young children was negatively associated with O5 retention for both male and female officers—both male and female officers with young children were more likely to leave. But the relationship is much stronger (and more negative) for female officers; female officers with young children were far more likely to leave as an O5 than were male officers, and this difference contributed positively to the gender gap in O5 retention for male and female officers. Interestingly, that contribution is relatively small in magnitude (less than one percentage point). The findings for the two promotion milestones—O4 and O6—are similar. Among those with very young children, men had higher promotion rates, while women had lower promotion rates relative to those without very young children, and these differences contributed positively to the gender gap in O4 and O6 promotion.

As for the role of preschool-age children (four to six), school-age children (seven to 18), and adult children, we find both male and female officers with children in these age groups were more likely to stay as an O3 relative to those without children in these age groups, and the relationship is stronger for female than male officers. This difference contributed to a narrowing of the gender gap in O3 retention. That is, the presence of children in these older age groups was advantageous to female officers in the sense that it reduced the retention rate gap between male and female officers that was attributable to the association component of the gap. As for the other milestones—promotion to O4, retention as an O5, and promotion to O6—we generally do not find that the contributions of having children in these older age groups are statistically significant.

Differences in coefficient estimates for the occupational group variables and, related, cumulative months deployed contributed to the part of the gender gap in retention and promotion rates attributable to the association component. Both male and female officers in tactical occupational specialties had a higher O3 retention rate, with the estimated relationship stronger for female than for male officers. The stronger relationship narrowed the gender gap in O3 retention. Related to this finding, the relationship between additional months of deployment and O3 retention is positive and statistically significant for female officers but negative and statistically significant for male officers. This difference in the association between deployment and O3 retention also narrowed the gender gap in O3 retention; it contributed negatively to the association component of the gender gap. Another notable finding with respect to deployment time is that cumulative months of deployment were positively associated with O6 promotion, for both male and female officers, but there was a stronger relationship for female officers. Consequently, months of deployment were advantageous to female officers, with respect to the gender gap in O6 promotion that is attributable to the association component.

Results for Specific Milestones

We next turn to the specific results at each milestone. We begin with retention as an O3 and the O4 promotion window (conditional on being promoted to O3), though, as shown in Figure 4.1, association differences are relatively unimportant for this milestone, accounting for 3.9 out of the 11.9-percentage-point difference in the likelihood of reaching this milestone for men and women. Figure 4.9 shows the decomposition of the component because of differences in associations, though the individual contributions do not sum to 3.9 percentage points, because we omit the contribution associated with the difference in the constant term. The figure shows that several factors contributed to this component, though family status is clearly particularly important. Family status was positively related to retention as an O3 for men (conditional on promotion to O3) but negatively associated with retention for female officers. Thus, a married female officer was less likely to stay, while a married male officer was more likely to stay.

Figure 4.9
Probability of Retention as an O3: Decomposition of the Part of the Gender Difference Attributable to the Association Between Outcome and Characteristics

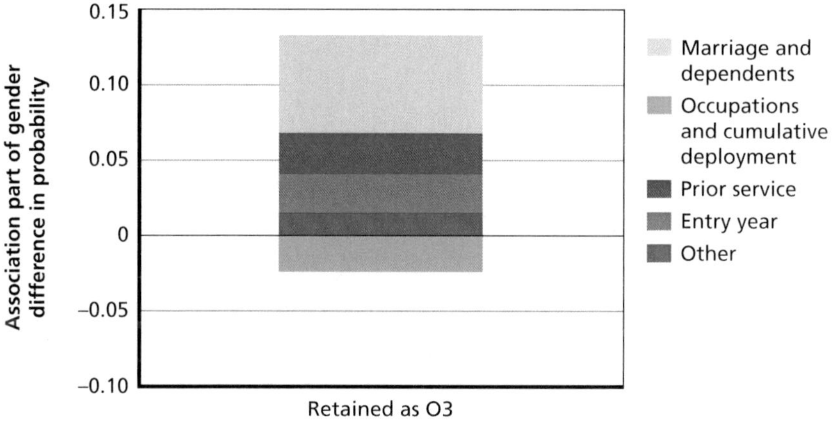

NOTE: The "other" category includes the contributions of commissioning source, race, ethnicity, service branch, and education.
RAND *RR1288-4.9*

Figures 4.10 to 4.12 show similar decompositions of the association component for promotion to O4 (conditional on retention as an O3), retention as an O5 (conditional on promotion to O5), and promotion to O6 (conditional on retention as an O5). As in Figure 4.9, the decompositions do not show the contribution of the difference in constant terms. For the two promotion milestones—promotion to O4 and to O6—family factors and race and ethnicity are key contributors to the gender difference because of differences in the associations between characteristics and these two outcomes. Being married is positively associated with promotion to O4 for male officers (conditional on retention as an O3) and with promotion to O6 (conditional on retention as an O5). But the association between marital status and promotion to either O4 or O6 is not statistically different from zero for female officers. Regarding the result for race, we find that white male officers had a higher O4 promotion rate relative to other groups, while the O4 promotion rate for white female officers is not statistically

Figure 4.10
Probability of Promotion to O4: Decomposition of the Part of the Gender Difference Attributable to the Association Between Outcome and Characteristics

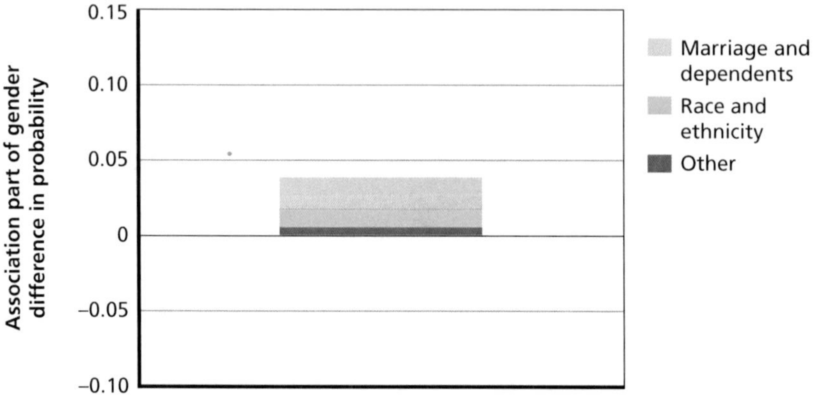

NOTE: The "other" category includes the contributions of prior service, occupations, commissioning source, entry year, service branch, cumulative deployment, and education.
RAND *RR1288-4.10*

significant. This difference in the association between being white and O4 promotion contributed positively to the gender gap in O4 promotion. It is important to note that in the case of promotion to O4, association differences were relatively unimportant in general, and, in the case of O6, cumulative deployment experience was also a contributor.

Figure 4.1 showed that for retention as an O5, the association component is a major share of the overall gender difference in the likelihood of reaching this milestone. We find that almost none of the characteristics had a statistically significant contribution to this component; the constant term was the main contributor that was statistically significant. This means that "being female" relative to "being male" was the main contributor in terms of statistical significance. Thus, other than entry year and prior service status, we are unable to decompose the association component, given the characteristics we include in our analysis.

Given the role of marital status as a contributor to the part of the gender differences in promotion to O4 and O6 and in retention as an

Figure 4.11
Probability of Retention as an O5: Decomposition of the Part of the Gender Difference Attributable to the Association Between Outcome and Characteristics

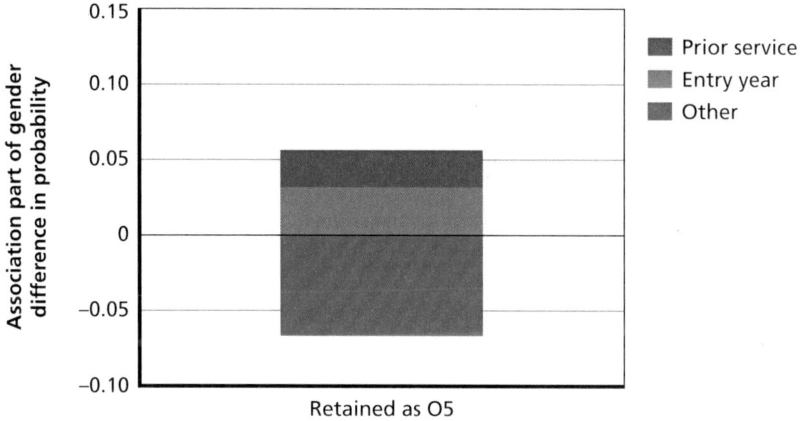

NOTE: The "other" category includes the contributions of family status, race and ethnicity, occupations, commissioning source, service branch, cumulative deployment, and education.
RAND *RR1288-4.11*

O3, attributable to differences in associations, we further decompose the contributions of marital status. The additional decomposition is shown in Figure 4.13. We find being a dual-military spouse contributed little to the association component. That said, unlike the part of the gender difference attributable to differences in observed characteristics (shown in Figure 4.9), being a dual-military spouse contributed positively to the gender difference (i.e., it is disadvantageous to female officers), but in none of these cases was the contribution statistically different from zero.

Decomposing the Gender Gap into Contributions of Specific Characteristics

So far, we have discussed the contribution of specific characteristics to the observed and the association components of the gender gap. We conclude this chapter by summing the contributions and showing the

Figure 4.12
Probability of Promotion to O6: Decomposition of the Part of the Gender Difference Attributable to the Association Between Outcome and Characteristics

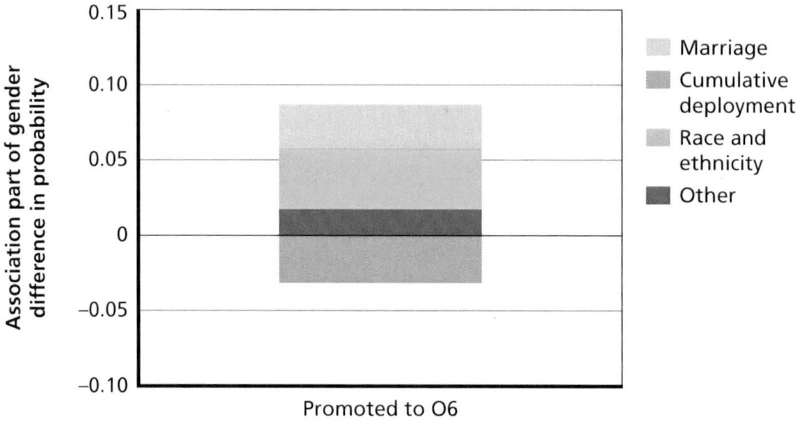

NOTE: The "other" category includes the contributions of dependents, prior service, occupations, commissioning source, entry year, service branch, and education.
RAND *RR1288-4.12*

Figure 4.13
Probability of Achieving Key Career Milestones: Decomposition of Gender Difference in the Association Component and Part Attributable to Dual-Military Status

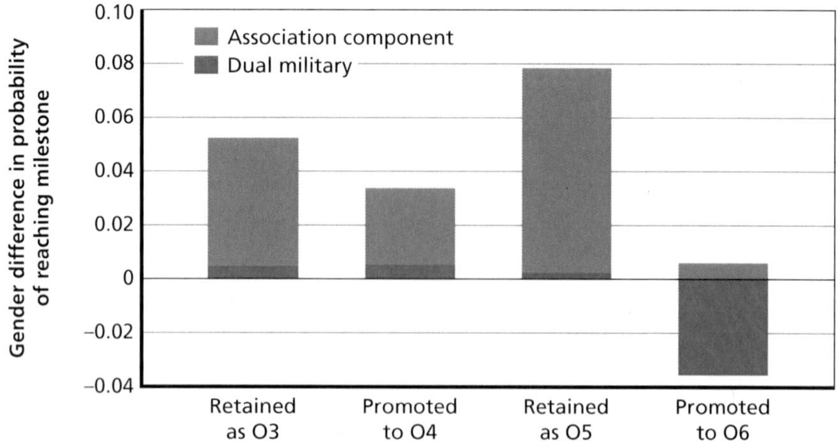

RAND *RR1288-4.13*

total contribution of specific characteristics to the gender gap in the probabilities of reaching different career milestones. We focus the discussion on the three milestones with the larger gender gaps (shown in Table 3.1): retention as an O3, promotion to O4, and retention as an O5.

With respect to family status, we find that marital status and dependent status contributed positively to the gender gap in the probability of being retained as an O3 (Figure 4.14) and of being promoted to O4 (Figure 4.15), with no statistically significant effects in general for the probability of being retained as an O5 (Figure 4.15). Focusing on the results for marital status, being married was positively associated with retention as an O3 and promotion to O4. Female officers were also less likely to be married. At the same time, we find that among those who were married, male officers were more likely to be retained as an O3 and promoted to O4, and female officers were less likely to be retained as an O3 and had no statistically different promotion to O4 relative to unmarried female officers. Thus, both differences in observed marital rates and differences in the association between mar-

Figure 4.14
Probability of Retention as an O3: Decomposition of the Gender Difference Attributable to Specific Characteristics

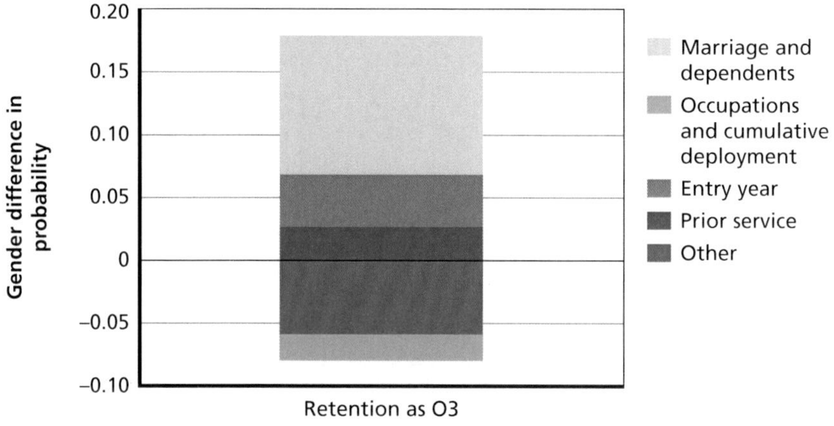

NOTE: The "other" category includes the contributions of commissioning source, service branch, education, race and ethnicity, and the constant term.
RAND *RR1288-4.14*

Figure 4.15
Probability of Promotion to O4: Decomposition of the Gender Difference Attributable to Specific Characteristics

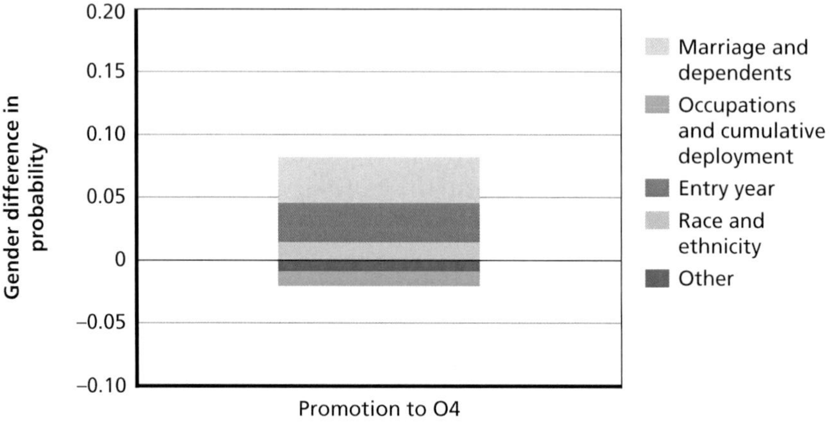

NOTE: The "other" category includes the contributions of being prior service, commissioning source, service branch, education, and the constant term.
RAND *RR1288-4.15*

riage and the probability of O3 retention and O4 promotion contributed to the positive gender gap.

The results for dependent status are similar in the sense that gender differences in the incidence of dependents and in the association between dependents status and the probability of O3 retention and O4 promotion contributed positively to the gender gap. For example, female officers were less likely to have very young children (birth to three), and the association between having young children and retention as an O3 was weaker for women than men. In the case of promotion to O4, male officers with very young children were more likely to be promoted, while female officers were less likely.

Turning to the results for occupational group, we combine the contribution of occupational group with that of cumulative months of deployment, because the two variables are related insofar as amount of deployment reflects occupational distribution within a given occupational grouping. Earlier, we found that the association component narrowed the gender gap in the probability of being retained as an O3, while the observed component increased the gender gap; the two components work in opposite directions. The overall contribution of occupational group and cumulative months of deployment to the gender gap in O3 retention is negative, but relatively small in magnitude (Figure 4.14). Female officers were less likely to be in tactical occupations, contributing positively to the observed component of the gender gap in O3 retention, but O3 retention was higher among officers in tactical occupations, with the association stronger for female officers. The overall contribution of being in a tactical occupation to O3 retention is negative but small in magnitude. The overall contribution of occupational group and cumulative months of deployment in O4 promotion is also negative. In this case, the key driver for this result is our finding that being in a tactical occupation is negatively associated with O4 promotion (see Figure 4.5).

As for retention as an O5, the contribution of occupational group and cumulative months of deployment operated in the same direction; these variables contributed positively to both the observed and association components of the gender gap. Female officers were less likely to be in tactical occupations and had fewer cumulative months of deployment, and both variables are positively associated with retention as an O5. At the same time, both variables are positively associated with O5 retention for male officers but are not statistically different from zero for female officers, also contributing to a positive gender gap in O5 retention. Taken together, the total effect is also positive (Figure 4.16).

In the case of O4 promotion, we find that race and ethnicity contributed positively to the gender gap (Figure 4.15). Here, the main driver was not differences in the contribution to the observed component. Instead, the main driver was the differences in the contribution to the association component. Specifically, as discussed in the context of Figure 4.10, white male officers had a higher O4 promotion rate relative to other groups, while the O4 promotion rate for white female officers is not statistically significant. As a result, being white was a positive contributor to the gender gap in O4 promotion.

Finally, entry year is another variable that contributed to the gender gap in career progression; specifically, it had an overall positive contribution to the gender gap in the probability of being retained as

Figure 4.16
Probability of Retention as an O5: Decomposition of the Gender Difference Attributable to Specific Characteristics

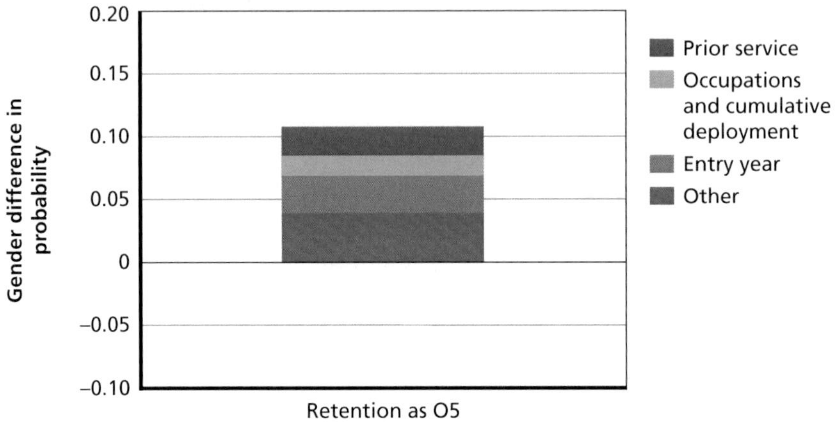

NOTE: The "other" category includes the contributions of marital status; having dependents ages zero to three, four to six, seven to 18, and over 18; commissioning source; service branch; education; and the constant term.
RAND *RR1288-4.16*

an O3, promoted to O4, and retained as an O5. For example, in the case of O3 retention, entry year contributed 4.1 percentage points to the overall 11.8-percentage-point difference between male and female officers in the probability of staying as an O3. Female officers were less represented in the earlier entry years and more represented in the later entry years. Entering in the early years was positively associated with O3 retention, while entering in the later years was negatively associated with O3 retention. Thus, observed differences in entry year contributed to the gender gap. Differences in associations also contributed to the gender gap in a disadvantageous way. Female officers who entered in earlier years had higher O3 retention than male officers, while female officers who entered in the later years had lower O3 retention than their male counterparts. Thus, both the association component and the observed component of the entry year are relevant to the gender gap at O3 retention.

CHAPTER FIVE

Concluding Thoughts

The largest differences between male and female officers' career progression occurred in the probabilities of being retained as an O3, being promoted to O4, and being retained as an O5. Our study of the factors explaining gender differences in the officer career pipeline, focusing on these milestones, leads to the following findings:

- The observed differences in job and individual characteristics between male and female officers explain some but not all of the differences in the likelihood of reaching retention and promotion milestones. At most, observed differences explain about two-thirds of the gender difference, with the remaining differences being attributable to differences in the association of characteristics and retention and promotion outcomes.
- Among the differences in observed characteristics, family factors—specifically, marital status and age and presence of dependents—are consistently important. Women's lower likelihood of being in a tactical occupation, lower levels of deployment experience, and increased likelihood to have entered in a recent cohort are also notable contributors to gender differences in the career pipeline. That is, family factors, occupational and deployment experience, and entry cohort tended to be major contributors toward explaining why women were less likely to be retained and promoted over their officer careers.
- Among the differences in the association between characteristics and outcomes, family factors had some role, but so did other characteristics.

- Overall, considering both the observed component and association component of the gender gap, marital status and dependents status are generally major positive contributors to gender differences in officer career progression. Both the observed component and association component played a role in the contribution of family factors.
- Among married officers, being a dual-military spouse appears to have had little or no role in contributing to gender differences in career progression.
- Overall, considering both the observed component and association component of the gender gap, occupational group contributed negatively to the gender gap in the midcareer (O3 retention and O4 promotion) but positively later in the career (O5 retention). Although female officers were less likely to be in tactical occupations, the negative contribution occurred because the role of that difference was outweighed by the negative contribution attributable to differences in the association between retention and promotion and being in a tactical occupation.
- Entry year contributed to the gender gap. Female officers were more likely to be represented in recent cohorts, and the association between entry year and retention and promotion differed between male and female officers in a way that led to lower retention and promotion rates for female officers.

Our study is descriptive and did not evaluate why these observed differences in characteristics occurred or why these characteristics' effects on retention and promotion outcomes might differ for male and female officers. Furthermore, our study did not evaluate any existing policies aimed at reducing differences in characteristics, such as opening occupations to female officers, or at reducing gender differences in career progression. That said, our study does suggest several implications with respect to policies aimed at closing the gender gap in officer career progression.

Given that observed differences in the characteristics of male and female officers contributed to differences in career progression, policies aimed at reducing those differences will contribute to narrowing the gender gap in career progression. For example, policies that increase the representation of female officers in tactical occupations, such as the 2013 repeal of the policy banning women from serving in infantry, artillery, armor, and other tactical occupations at the battalion size or smaller, will reduce occupational differences between male and female officers and contribute to reducing the part of the gender gap in the retention and promotion outcomes attributable to observed differences. Interestingly, our finding that dual-military officers exhibited similar patterns of career progression to other officers suggests that further efforts to address issues that are unique to dual-military officers might have little impact on reducing the gender gap in officer career progression.

While much of the recent policy focus has been on improving gender equality with respect to roles in combat and in all military occupations, our analysis suggests that, among policies that aim to reduce observed differences in characteristics, policies that target work-family balance could have an important impact on reducing the explained gender gap in career progression. DoD currently has some policies in place. For example, a policy allows a four-month deferment of an overseas assignment after the birth of a child, and the Career Intermission Pilot Program provides a one-time temporary transition from active duty to the Individual Ready Reserve for one to three years to pursue personal or professional obligations, including childbirth. The Air Force recently indicated that it is considering longer maternity leave, similar to changes recently announced by the Navy that would extend maternity leave to 18 weeks ("Air Force Considering Longer Maternity Leave," 2015). Our analysis does not indicate which policies that improve work-life balance are effective or would improve the gender gap in officer career progression. Instead, the analysis points to this area as one that deserves closer attention in the future, given the role of family factors in explaining the gender gap in career progression.

Still, our analysis indicates that policies that reduce observed differences in characteristics are unlikely to fully eliminate the gender gap in career progression. Differences in the association between specific factors and retention and promotion outcomes also play a role; for some career milestones, these differences were the major contributor to gender differences. The associations we estimated captured structural factors (e.g., the factors that cause retention and promotion outcomes to differ for male and female officers with the same observed characteristics), as well as the role of self-selection and endogeneity of both observed and unobserved characteristics. With respect to unobserved characteristics, gender differences in taste for the military and performance could be important. For example, taste for the military might be influenced by perceptions about the climate and acceptance of women in service, as well as ongoing issues related to sexual harassment and assault. Thus, policies aimed at reducing gender differences must also tackle differences in structural factors and differences in factors that are more difficult to measure.

In short, our analysis indicates that gender differences in career progression can be partly explained by differences in job-related characteristics, such as occupational differences and deployment experience, as well as individual characteristics, such as marital status and age and presence of dependents. Policies that reduce these differences will contribute to reducing the gender gap in officer career progression but will not eliminate it. Additional attention must be given to structural factors, including how retention decisions and the promotion process differ for male and female officers with the same characteristics, and to differences in factors that are more difficult to observe, such as differences in attitudes toward military service and performance.

APPENDIX A

Blinder-Oaxaca Methodology

The Blinder-Oaxaca method begins with two groups (in our case, male and female officers) and an outcome variable for each group (in our case, a retention or promotion outcome). We denote *group* with the subscript g (where g equals either male officers or female officers), and we denote the outcome for each group as Y_g.[1] The outcome for each group is assumed to be linearly related to a set of observed characteristics, X_g, where β_g is the slope parameters, and unobserved characteristics are v_g: $Y_g = X_g \beta_g + v_g$, for g = male, female. We assume that the unobserved and observed characteristics are independent: $E(v_g|X_g) = 0$, where $E()$ is the expectation operator, so that we can decompose the difference in mean outcome between groups into the explained and structural components.[2] We also assume $E(v_g) = 0$. The difference in the mean outcome between groups, R, is given by:

$$R = E(Y_m) - E(Y_f) = \bar{Y}_m = - \bar{Y}_f = E(X_m)\beta_m - E(X_f)\beta_f = \bar{X}_m \beta_m - \bar{X} =_f \beta_f.$$

We simplified the notation and replaced expected values with sample averages—e.g., we replaced $E(Y_m)$ with $\bar{Y}_m$. Through simple

1 We have ten retention and promotion outcomes to consider, as shown in Table 2.1, so we should also subscript each outcome by i, where i varies from one to ten. However, to reduce the clutter of notation, we drop the i subscript here and refer to a single outcome, though it should be recalled throughout that we have ten outcomes to consider. Later in this appendix, we discuss selectivity bias and how a given outcome later in the career can depend on the sample selection that occurs earlier in the career.

2 Later in this appendix, we discuss what happens when this assumption is violated, as is likely the case in our analysis, and the implications for interpretation of the results.

algebraic manipulation, adding and subtracting terms, R can be expressed as

$$R = \bar{X}_m\beta_m - \bar{X}_f\beta_f = (\bar{X}_m - \bar{X}_f)\beta_f + \bar{X}_m(\beta_m - \beta_f) = E + S.$$

The first term, $(\bar{X}_m - \bar{X}_f)\beta_f$, is the composition portion, or the explained portion, of the difference in the mean outcome because it represents the contribution of group differences in characteristics to the overall difference in the mean outcome. We denote this term as E. The second term is the contribution of the differences in the coefficients (including differences in the intercept), $\bar{X}_m(\beta_m - \beta_f)$. Insofar as the coefficients capture structural factors in the retention decision process and in the promotion process that translates characteristics into outcomes, this term is often termed the *structural portion* of the difference in mean outcomes.[3]

The expression for S presupposes that the appropriate benchmark for the female coefficients is the male coefficients. The expression implicitly assumes that the assessment of the structure of the retention decision and promotion processes for female officers should be based on how it compares with those processes for male officers. However, using the coefficients for male officers as the benchmark could be problematic. For example, in the literature on labor market discrimination, researchers have argued that using the male coefficients as the reference group in an Blinder-Oaxaca decomposition could overstate the degree of discrimination against women if some of the observed differences between male and female labor market outcomes are due to discrimination (possibly positive discrimination) directed toward men and in any case, the choice of reference group (either female or male) yields different results (Cotton, 1988; Oaxaca and Ransom, 1994). As we will discuss, we do not argue that the term S measures discrimination. Nor do we argue that S has a structural interpretation, as in the

[3] The decomposition for R is not unique. An alternative expression for R is $R = (\bar{X}_m - \bar{X}_f)\beta_m + \bar{X}_f(\beta_m - \beta_f)$. Here, the differences in characteristics are weighted by the coefficients for male officers, rather than female officers, while the differences in coefficients are weighted by the characteristics for female officers, rather for male officers. The choice of which to use is arbitrary.

labor market discrimination literature. Still, using the coefficients for male officers as the benchmark is somewhat arbitrary, so we consider an alternative proposed in the literature.

Following Neumark (1988) and Fortin (2007), we use the coefficients from a pooled regression as the benchmark.[4] The pooled regression is $Y = X\beta^* + v$.

Again, using simple algebra, we can express R as

$$R = (\bar{X}_m - \bar{X}_f)\beta^* + \bar{X}_m(\beta_m - \beta^*) - \bar{X}_f(\beta_f - \beta^*) = E^* + S^*.$$

Here, the explained component of the decomposition that accounts for group differences in characteristics, E^* is weighted by β^*, while the structural component, S^*, has two elements. The first is related to the difference between the male coefficients and the benchmark, and the second is related to the difference between the female coefficients and the benchmark:

$$U^* = \bar{X}_m(\beta_m - \beta^*) - \bar{X}_f(\beta_f - \beta^*).$$

In the literature, the first structural component is called the *advantage to men* $[\bar{X}_m(\beta_m - \beta^*)]$, while the second structural component is called the *disadvantage to women* $[-\bar{X}_f(\beta_f - \beta^*)]$.

In Chapter Four, we presented estimates of E^* and S^* for each career milestone. These estimates provided information on the overall explained and structural component of differences in mean outcome between male and female officers for each milestone. As we discuss below, we interpret S^* as the unexplained component rather than the structural component.

We can further decompose E^* and S^* to get the detailed contribution of each characteristic to the explained and structural compo-

---

4 An issue discussed in the literature is that the use of pooled regression estimates without a dummy variable for group membership (e.g., an indicator for female officers) can inappropriately attribute some of the unexplained component to the explained component. This is a classic omitted-variable problem. We follow the approach taken in the literature (see Jann, 2008) and include a female-officer indicator in the pool-regression models as an additional covariate.

nents. The individual contribution of each characteristic is straightforward because the total component is a simple sum over the individual contributions of each characteristic. For example, consider E^*. It can be decomposed to the individual characteristics as follows:

$$E^* = (\bar{X}_m - \bar{X}_f)\beta^* = (\bar{X}_m^1 - \bar{X}_{mf}^1)\beta_1^* + (\bar{X}_m^2 - \bar{X}_{mf}^2)\beta_2^* + \ldots,$$

where $\bar{X}^1$, $\bar{X}^2$, . . . are the means of the single characteristics, and β_1^*, β_2^*, . . . are the associated coefficients. The first element in the right-hand side is the contribution of the group differences in $\bar{X}^1$ and so forth. A similar detailed decomposition can be made of S^*. In Chapter Four, we presented both the overall and the detailed decompositions for milestones with the largest differences in outcomes.

Issues

There are two issues we must address regarding performing the Blinder-Oaxaca decomposition in our study. The first has to do problems with performing detailed decompositions with indicator variables when the choice of the omitted category is arbitrary. We address this issue by using a normalization procedure discussed in Yun (2005) and summarized in Jann (2008) and Fortin (2007). The second issue has to do with self-selection and endogeneity, which imply a violation of a key assumption, $E(v|X) = 0$.

The Problem of the Choice of the Omitted Group for Categorical Variables

The first issue arises because most of the characteristics we consider in our regression analysis, such as occupational group, are categorical, or indicator, variables. As is well-known in the literature, one category is omitted in linear regression models to avoid perfect collinearity with the intercept. For example, we estimate a pooled regression model with indicator variables for each occupational group, with tactical occupations as the omitted category. The problem for the Blinder-Oaxaca decomposition is that the contributions to the structural part, S^*, of the decomposition of these indicator variables will arbitrarily depend on which category is omitted. For example, in our analysis, if we choose a different occupational group as the omitted group—e.g., administrative occupations—the contribution of each occupation to the structural portion differs from when we choose tactical occupations as the omitted group. Since the choice of which group to omit is arbitrary, the detailed decomposition, which changes when the omitted group is changed, is arbitrary as well. It is important to note that the detailed decomposition of the explained part is unaffected by this problem.

We handle this issue by transforming the estimates of the categorical variables for a given characteristic so the decomposition is independent of the choice of omitted category. For example, suppose the characteristic is occupational group and we have six groups to consider, with one as the omitted category. First, given that the coefficient on the omitted category is zero, by definition, we take the average of the coefficient estimates for the characteristic, (including the zero estimate for the omitted category), where the average is given by c. Second, we then add c to the intercept and subtract c from each coefficient estimate, including the zero coefficient for the omitted category. Thus, the coefficients are expressed as deviations from the overall mean of the coefficients across all subcategories. The results are that the coefficient estimates (including the omitted category) sum to zero, and the transformed model is equivalent to the untransformed model. These transformed coefficient estimates are then used in the decomposition. This approach yields the same detailed decomposition regardless of which category is omitted. The approach is described by Yun (2005) and Fortin (2007), and we make use of the DEVCON program created by Jann (2005) in Stata to implement the approach.

The Problem of Self-Selection and Endogeneity

We face two issues that result in the violation of the independence assumption $[E(v|X) = 0]$. The first issue is self-selection and the problem that, at any given career milestone, the group of officers eligible to proceed to the next milestone (i.e., the group that participates in a particular milestone) is a self-selected group of personnel. The second issue is that our data do not include unobserved factors (by definition), and these factors, as well as observed factors, could be correlated with the factors that we do observe. We discuss each issue in turn.

Self-selection is particularly a problem when the self-selection process differs for male and female officers and might be partially based on factors that are unobservable in our data. For example, those eligible for promotion to O4 are those who were promoted to O3 and stayed until the promotion eligibility window. The eligible group is not a random group of officers but consists of those who were selected for earlier promotions and who made retention decisions and opted to stay. These officers were likely to be personnel who stayed because they believed that they have a good chance of a future promotion, while those who thought that they had a poorer chance of future promotion were more likely to leave. Furthermore, those selected for promotion at earlier stages of the career pipeline were likely to be those who were viewed as having good career prospects. Consequently, the eligibility pool consisted of those who self-selected into the pool. Importantly, this self-selection might differ for male and female officers and is likely to be partially based on factors that are not observable in our data, such as ability and performance.

When participation in the eligibility pool for a particular milestone is based on unobserved factors, the expected value of a given outcome is no longer $\bar{X}_g\beta_g$ but is instead given by $E(Y) = \bar{X}\beta^* + E(v|eligibility)$. This expression can be rewritten as $E(Y) = \bar{X}\beta^* + \bar{\lambda}\sigma^*$, where $\bar{\lambda}$ is the mean of the usual inverse Mills ratio term that is obtained by estimating a probit model of the likelihood of eligibility, and σ^* is the coefficient on the Mills ratio term in a pooled regression, related to the variance of the error term of the selection eligibility process and the covariance between that error and v. Performing standard ordinary least squares regression without accounting for the term $\bar{\lambda}\sigma^*$ will lead to biased estimates of β^* if the eligibility selection process is correlated with the outcome of interest. This is likely to be the case in our analysis. Selection bias will also affect the decomposition. The Blinder-Oaxaca decomposition becomes $R = (\bar{X}_m - \bar{X}_f)\beta^* + \bar{X}_m(\beta_m - \beta^*) - \bar{X}_f(\beta_f - \beta^*) + (\bar{\lambda}_m - \bar{\lambda}_f)\sigma^* + \bar{\lambda}_m(\sigma_m - \sigma^*) - \bar{\lambda}_f(\sigma_f - \sigma^*) = E^* + S'$.

E^* is defined as before, but the structural portion now includes not only $\bar{X}_m(\beta_m - \beta^*) - \bar{X}_f(\beta_f - \beta^*)$ but additional terms related to differences attributable to unobserved characteristics, $(\bar{\lambda}_m - \bar{\lambda}_f)\sigma^*$, as well as differences in the effects of those unobservables (e.g., the coefficients on the unobservables), $\bar{\lambda}_m(\sigma_m - \sigma^*) - \bar{\lambda}_f(\sigma_f - \sigma^*)$. If the self-selection processes differs for men and women, the additional terms in S' do not disappear.

One approach to addressing this issue is to use a correction procedure, such as the one recommended by Heckman (1974, 1976), though Neuman and Oaxaca (2004) argued that the Heckman approach introduces fundamental ambiguities about how to interpret the structural component of the decomposition. However, even if valid, these procedures require an instrumental variable that predicts eligibility but not the outcome of interest. That is, there is an exclusion criterion.

Unfortunately, we lack such an instrument, given the state of the data we have available. Consequently, we cannot give a causal interpretation to our results. The explained portion and the structural portion of the decomposition of the gender differences in the officer career pipeline will reflect factors that affected the likelihood of reaching a given milestone *and* the selection factors that resulted in an officer being eligible to reach that milestone. Furthermore, because S' also includes differences attributable to unobserved characteristics and differences in the selection processes, we can no longer interpret this part as structural. Instead, we call this the *unexplained part* of the decomposition because it includes both the structural part—how a given characteristic affects the gender gap in reaching a given milestone—and the differential selection process that leads male and female officers to be eligible to reach that milestone.

Thus, we are unable to state that a given observed difference in male and female characteristics cause a given gender gap in the career pipeline or state that unexplained differences are due to structural differences and possibly discrimination in how men and women are treated in the military. Instead, we can only state the factors that contribute to the gender gap when the contribution reflects both causal factors and selection factors.

The second issue we must confront is endogeneity of the characteristics. With respect to the unobserved characteristics, we assume that the observed and unobserved characteristics are independent, but it is possible that they are correlated. For example, having an advanced degree (observed) might be correlated with taste for the military or with ability (unobserved). If so, and we did not account for this correlation, the coefficient estimates would be biased.

Similarly, observed characteristics, especially those related to family formation, could be endogenous. Individuals might choose the timing of marriage and childbearing based on their expectations about current and future promotions, as well as retention plans. Officers might marry, have children, and then decide whether to stay or leave. Alternatively, they might leave because they plan to marry or have children or they might defer marriage or childbearing until completing an obligation or achieving a promotion. As with the case of unobserved characteristics and self-selection, not accounting for this endogeneity could result in biased coefficient estimates.

Given that we have no means to correct for this issue, we must use caution in interpreting our results as causal effects; acknowledge that the decompositions we compute can reflect correlations between observed and unobserved characteristics; and only argue that our analysis shows the contributions of a given factor, where that contribution reflects both causal relationships and possible correlation with observed and unobserved factors.

In sum, we decompose observed differences in retention and promotion outcomes for male and female officers into a part attributable to differences in observed characteristics and a part attributable to differences in the associations between these characteristics and the outcomes. The second part shows the differences in outcomes for male and female officers with the same observed characteristics. Because we cannot give a causal interpretation to the estimated regression coefficients—because of self-selection and endogeneity related to both observed and unobserved characteristics—the second part does not show the differences in the effects of the observed characteristics, just the differences in their associations.

APPENDIX B

Detailed Results

Chapter Four presents an overview of the results of our analysis. This appendix provides detailed results.

Table B.1 shows the descriptive statistics for all of the variables included in our regression analysis and summarized in Chapter Three. Table B.2 shows the estimated parameters of the pooled regression at each milestone; the coefficient estimates are normalized so that the effects are expressed as deviations from the grand mean. Table B.3 shows, at each milestone, the Blinder–Oaxaca decomposition into the explained and unexplained components and the further detailed decomposition of the explained component.

Table B.1
Means of Variables at Each Career Milestone

Variable	Entry	Retained as O1	Promoted to O2	Retained as O2	Promoted to O3	Retained as O3	Promoted to O4	Retained as O4	Promoted to O5	Retained as O5
Female (%)	14.9	14.9	14.8	14.8	10.5	9.1	8.5	8.4	9.3	8.3
Retained/ promoted (%)										
All	99.9	98.0	99.5	92.7	82.0	83.6	92.3	85.4	86.2	59.2
Male	99.9	98.2	99.6	93.2	83.2	84.1	92.4	85.6	87.2	59.3
Female	99.8	97.2	99.2	89.6	71.6	78.1	91.3	83.5	76.3	58.0
Education (%)										
Bachelor's degree										
All	94.6	94.6	95.7	95.8	93.8	93.7	54.9	53.9	24.5	24.6
Male	94.3	94.3	95.7	95.7	94.1	94.0	56.2	55.3	25.3	25.3
Female	96.2	96.2	96.1	96.1	91.1	90.5	41.1	39.2	16.7	16.8
More than bachelor's degree										
All	5.4	5.4	4.3	4.2	6.2	6.3	45.1	46.1	75.5	75.4
Male	5.7	5.7	4.3	4.3	5.9	6.0	43.8	44.7	74.7	74.7

Table B.1—Continued

Variable	Entry	Retained as O1	Promoted to O2	Retained as O2	Promoted to O3	Retained as O3	Promoted to O4	Retained as O4	Promoted to O5	Retained as O5
Female	3.8	3.8	3.9	3.9	8.9	9.5	58.9	60.8	83.3	83.2
DoD occupational group (%)										
Tactical										
All	24.4	24.4	36.2	36.2	54.7	56.0	58.8	57.9	57.4	59.2
Male	27.0	27.0	39.4	39.4	57.8	58.8	62.7	61.9	62.1	63.4
Female	9.0	9.0	18.0	18.0	27.5	27.8	16.5	15.0	12.2	12.4
Intelligence										
All	4.7	4.7	5.9	5.9	7.5	7.2	7.5	7.6	6.6	6.1
Male	3.6	3.6	4.9	4.9	7.0	6.8	7.0	7.1	6.2	5.8
Female	10.7	10.7	11.4	11.5	12.0	11.1	12.2	12.1	10.3	10.0
Engineering										
All	16.9	16.9	18.8	18.8	15.6	15.3	13.6	13.9	13.5	13.4
Male	16.0	16.0	18.0	18.1	15.1	14.9	12.9	13.1	13.3	13.2
Female	22.2	22.3	22.9	23.0	20.1	19.8	21.7	22.3	15.9	16.5

Table B.1—Continued

Variable	Entry	Retained as O1	Promoted to O2	Retained as O2	Promoted to O3	Retained as O3	Promoted to O4	Retained as O4	Promoted to O5	Retained as O5
Administration										
All	7.8	7.8	8.9	8.9	8.1	7.8	7.9	8.0	9.2	8.6
Male	6.3	6.3	7.2	7.2	6.5	6.4	5.7	5.8	5.6	5.4
Female	16.5	16.5	18.5	18.5	21.2	22.5	31.5	32.1	44.1	44.1
Supply/ procurement										
All	7.3	7.3	11.2	11.2	11.2	10.7	11.6	12.0	12.7	12.1
Male	6.1	6.1	10.1	10.1	10.5	10.1	11.0	11.4	12.3	11.7
Female	14.2	14.2	17.9	17.9	17.6	17.5	17.8	18.3	16.8	16.3
Other										
All	39.0	39.0	19.0	19.0	2.9	3.0	0.6	0.6	0.6	0.5
Male	41.0	41.0	20.4	20.3	3.1	3.1	0.7	0.7	0.6	0.5
Female	27.4	27.3	11.3	11.1	1.5	1.4	0.3	0.3	0.6	0.6
Cumulative deployment (months)										
All	0.09	2.50	2.2	6.9	4.7	14.9	11.5	19.3	12.8	20.1

Table B.1—Continued

Variable	Entry	Retained as O1	Promoted to O2	Retained as O2	Promoted to O3	Retained as O3	Promoted to O4	Retained as O4	Promoted to O5	Retained as O5
Male	0.09	2.48	2.2	7.0	4.7	15.2	11.9	20.0	13.6	21.0
Female	0.07	2.65	2.4	6.4	4.3	11.9	7.0	11.9	5.7	10.3
Prior enlisted service (%)										
All	28.5	28.5	23.1	23.1	22.0	24.3	18.2	17.6	14.7	13.1
Male	29.5	29.5	24.1	24.1	22.6	24.7	18.5	17.9	14.8	13.2
Female	22.7	22.7	17.3	17.3	16.3	20.0	15.1	14.3	13.3	12.1
Source of commission (%)										
Academy										
All	24.9	24.9	24.5	24.5	25.7	23.5	22.7	22.4	25.5	26.4
Male	24.9	24.9	24.8	24.8	25.9	23.8	23.0	22.9	26.4	27.2
Female	24.7	24.7	22.7	22.7	24.1	20.9	18.5	17.8	15.9	16.8
ROTC										
All	37.8	37.8	43.9	43.9	48.4	48.4	52.6	53.3	43.8	44.4
Male	36.3	36.3	42.4	42.4	47.8	48.0	52.3	53.0	44.5	44.9
Female	46.4	46.5	52.3	52.3	53.7	53.3	55.4	56.5	37.4	38.7

Table B.1—Continued

Variable	Entry	Retained as O1	Promoted to O2	Retained as O2	Promoted to O3	Retained as O3	Promoted to O4	Retained as O4	Promoted to O5	Retained as O5
OCS/OCT										
All	28.0	28.0	24.4	24.5	20.9	22.6	21.9	21.5	28.4	26.8
Male	28.8	28.8	25.0	25.0	21.2	22.7	21.6	21.2	26.6	25.3
Female	23.4	23.4	21.0	21.0	18.6	21.7	24.7	24.6	45.9	43.8
Other										
All	9.3	9.3	7.1	7.1	5.0	5.4	2.9	2.8	2.3	2.4
Male	10.0	10.0	7.7	7.7	5.1	5.5	3.0	3.0	2.5	2.6
Female	5.5	5.5	4.0	4.0	3.6	4.2	1.3	1.1	0.7	0.7
Service branch (%)										
Army										
All	28.8	28.8	31.4	31.5	25.2	23.0	25.6	26.7	21.2	20.0
Male	27.5	27.6	30.5	30.5	24.8	22.7	25.4	26.5	21.5	20.2
Female	35.8	35.9	36.9	37.1	28.5	26.5	28.2	28.9	19.1	17.9
Navy										
All	22.7	22.7	20.7	20.6	25.3	24.9	24.2	23.3	44.2	44.1
Male	23.3	23.3	21.4	21.4	25.7	25.3	24.0	23.1	43.6	43.6

Table B.1—Continued

Variable	Entry	Retained as O1	Promoted to O2	Retained as O2	Promoted to O3	Retained as O3	Promoted to O4	Retained as O4	Promoted to O5	Retained as O5
Female	19.1	19.1	16.4	16.2	21.8	21.6	26.5	25.3	49.6	49.7
Marines										
All	14.3	14.3	10.7	10.7	9.9	10.1	8.2	7.9	3.5	3.7
Male	15.4	15.4	11.5	11.5	10.5	10.7	8.7	8.5	3.8	3.9
Female	8.1	8.1	5.8	5.8	4.4	4.2	2.4	2.2	1.2	1.2
Air Force										
All	24.9	24.9	30.1	30.1	34.6	36.5	39.1	39.3	28.7	29.8
Male	23.7	23.7	28.9	28.9	33.8	35.8	38.8	39.0	28.7	29.7
Female	31.5	31.5	36.9	36.9	41.7	43.4	41.7	42.5	29.4	30.5
Race/ethnicity (%)										
White										
All	75.5	75.5	76.9	77.0	81.1	81.0	85.7	85.4	89.0	89.1
Male	77.3	77.3	78.7	78.7	82.2	82.2	86.7	86.4	89.8	89.8
Female	65.1	65.2	67.0	67.0	71.2	69.2	74.4	74.0	81.6	80.9
Black										
All	9.4	9.4	8.9	8.9	8.0	8.1	7.4	7.5	6.2	6.1

Table B.1—Continued

Variable	Entry	Retained as O1	Promoted to O2	Retained as O2	Promoted to O3	Retained as O3	Promoted to O4	Retained as O4	Promoted to O5	Retained as O5
Male	8.2	8.2	7.7	7.7	7.1	7.2	6.5	6.6	5.5	5.4
Female	16.6	16.6	15.8	15.8	15.2	17.2	16.3	16.8	12.5	13.3
Hispanic										
All	5.8	5.8	5.3	5.2	3.8	3.7	2.3	2.3	1.7	1.7
Male	5.7	5.7	5.2	5.1	3.8	3.7	2.3	2.2	1.6	1.6
Female	6.4	6.4	5.8	5.8	3.9	3.7	2.8	2.8	2.3	2.3
Asian/Pacific										
All	9.3	9.3	8.9	8.9	7.1	7.1	4.7	4.8	3.1	3.2
Male	8.9	8.9	8.4	8.4	6.8	6.9	4.5	4.7	3.0	3.1
Female	11.8	11.8	11.3	11.3	9.7	9.8	6.5	6.4	3.6	3.5
Marital status (%)										
Married										
All	46.8	46.8	57.6	57.7	79.8	83.7	91.0	91.3	92.7	93.1
Male	48.1	48.1	59.3	59.3	81.7	85.7	93.0	93.4	95.2	95.4
Female	39.0	39.1	47.9	47.9	62.9	63.7	69.5	69.0	68.5	67.6

Table B.1—Continued

Variable	Entry	Retained as O1	Promoted to O2	Retained as O2	Promoted to O3	Retained as O3	Promoted to O4	Retained as O4	Promoted to O5	Retained as O5
Married (not dual military)										
All	42.4	42.5	51.7	51.8	74.2	78.1	86.1	86.4	89.5	90.0
Male	45.6	45.6	55.7	55.8	78.2	81.9	89.4	89.8	92.6	92.8
Female	24.6	24.6	28.6	28.7	40.2	39.8	50.0	50.0	59.8	59.7
Married (dual military)										
All	4.3	4.3	5.9	5.9	5.5	5.6	4.9	4.9	3.2	3.0
Male	2.6	2.6	3.5	3.5	3.5	3.8	3.6	3.6	2.6	2.6
Female	14.5	14.4	19.3	19.3	22.7	23.9	19.5	19.0	8.7	7.8
Not married										
All	53.2	53.2	42.4	42.3	20.2	16.3	9.0	8.7	7.3	6.9
Male	51.9	51.9	40.7	40.7	18.3	14.3	7.0	6.6	4.8	4.6
Female	61.0	60.9	52.1	52.1	37.1	36.3	30.5	31.0	31.5	32.4

Table B.1—Continued

Variable	Entry	Retained as O1	Promoted to O2	Retained as O2	Promoted to O3	Retained as O3	Promoted to O4	Retained as O4	Promoted to O5	Retained as O5
Dependents (%)										
Any dependents (all ages)										
All	24.1	24.1	29.8	29.9	60.9	67.4	83.5	84.3	87.3	87.6
Male	25.4	25.4	31.2	31.3	63.4	69.6	85.9	86.8	90.5	90.5
Female	16.9	16.9	21.9	21.8	39.6	44.9	57.0	57.6	56.1	55.5
Any dependents ages 0–3										
All	13.6	13.7	19.0	19.0	37.5	40.2	26.6	25.8	9.2	8.7
Male	14.8	14.8	20.2	20.2	39.2	41.9	27.2	26.4	9.4	9.1
Female	6.8	6.8	12.3	12.3	22.6	23.4	19.6	18.8	6.4	4.3
Any dependents ages 4–6										
All	8.7	8.7	8.6	8.6	24.7	28.2	32.0	32.3	15.6	15.1
Male	9.3	9.3	9.3	9.4	26.2	29.6	33.3	33.5	16.0	15.5

Table B.1—Continued

Variable	Entry	Retained as O1	Promoted to O2	Retained as O2	Promoted to O3	Retained as O3	Promoted to O4	Retained as O4	Promoted to O5	Retained as O5
Female	4.8	4.8	4.1	4.0	11.4	14.1	18.9	19.5	11.3	10.0
Any dependents ages 7–18										
All	11.8	11.8	12.7	12.7	26.9	31.4	62.3	64.0	73.6	73.9
Male	12.3	12.3	13.2	13.2	28.2	32.5	64.8	66.5	77.0	76.8
Female	9.4	9.4	9.7	9.8	16.3	20.8	35.7	37.0	41.4	42.0
Any dependents ages 18+										
All	1.5	1.5	1.8	1.8	4.7	5.5	11.0	11.2	28.1	29.3
Male	1.4	1.4	1.7	1.7	4.5	5.2	10.9	11.1	29.2	30.4
Female	2.3	2.3	2.8	2.8	6.4	8.4	12.5	12.6	17.0	17.9
Joint spouse (%)										
All	4.5	4.5	5.9	5.9	5.5	5.6	4.9	4.9	3.2	3.0
Male	2.7	2.7	3.5	3.5	3.5	3.8	3.6	3.6	2.6	2.6
Female	15.3	15.3	19.3	19.3	22.7	23.9	19.5	19.0	8.7	7.8
Total observations	68,559	68,487	93,021	92,585	68,190	55,714	35,268	32,563	11,653	10,045

Table B.2
Pooled Regression Estimates, Male Officer Regression Estimates, and Female Officer Regression, Selected Milestones (Linear Probability Model with Normalized Categorical Variables)

	Probability Retained as O3			Probability Promoted to O4			Probability Retained as O5			Probability Promoted to O6		
Variable	Pooled	Males	Females	Pooled	Males	Females	Pooled	Males	Females	Pooled	Males	Females
Female indicator	−0.0396***			−0.030***			−0.0757***			0.0355*		
Cumulative months of deployment	−0.000448*	−0.000638**	0.00227**	0.0015***	0.00149***	0.00180***	0.0008***	0.00076**	0.00143	0.0038***	0.0036***	0.0063***
Married	0.0465***	0.0598***	−0.0120**	0.0202***	0.0258***	−0.00354	0.00948	0.0193**	−0.00316	0.0467***	0.0676***	0.0178
Not married	−0.0465***	−0.0598***	0.0120**	−0.020***	−0.0258***	0.00354	−0.00948	−0.0193**	0.00316	−0.0467***	−0.0676***	−0.0178
Any dependents ages 0–3	0.0275***	0.0265***	0.0165***	0.00321*	0.00428**	−0.0206***	−0.0215***	−0.0147***	−0.123***	0.0109	0.0156*	−0.0900**
No dependents ages 0–3	−0.0275***	−0.0265***	−0.0165***	−0.00321*	−0.00428**	0.0206***	0.0215***	0.0147***	0.123***	−0.0109	−0.0156*	0.0900**
Any dependents ages 4–6	0.0321***	0.0292***	0.0554***	0.0094***	0.00859***	0.00766	−0.0115***	−0.0107**	−0.0270	0.0140**	0.0154**	−0.00552
No dependents ages 4–6	−0.0321***	−0.0292***	−0.0554***	−0.009***	−0.00859***	−0.00766	0.0115***	0.0107**	0.0270	−0.0140**	−0.0154**	0.00552
Any dependents ages 7–18	0.0517***	0.0513***	0.0612***	0.0082***	0.00890***	0.00475	−0.00638*	−0.00826**	0.00632	0.00850	0.0144**	−0.0337*

Table B.2—Continued

Variable	Probability Retained as O3			Probability Promoted to O4			Probability Retained as O5			Probability Promoted to O6		
	Pooled	Males	Females	Pooled	Males	Females	Pooled	Males	Females	Pooled	Males	Females
No dependents ages 7–18	−0.0517***	−0.0513***	−0.0612***	−0.008***	−0.00890***	−0.00475	0.00638*	0.00826**	−0.00632	−0.00850	−0.0144**	0.0337*
Any dependents ages 19+	0.0440***	0.0396***	0.0634***	−0.013***	−0.0171***	0.0140	0.0278***	0.0287***	0.0138	−0.0459***	−0.0453***	−0.0228
No dependents ages 19+	−0.0440***	−0.0396***	−0.0634***	0.013***	0.0171***	−0.0140	−0.0278***	−0.0287***	−0.0138	0.0459***	0.0453***	0.0228
Prior service	0.0242***	0.0204***	0.0576***	0.026***	0.0249***	0.0366***	−0.0552***	−0.0581***	−0.0227	0.0104	0.00854	0.0397
No prior service	−0.0242***	−0.0204***	−0.0576***	−0.026***	−0.0249***	−0.0366***	0.0552***	0.0581***	0.0227	−0.0104	−0.00854	−0.0397
White	−0.0156***	−0.0125***	−0.0247***	0.009***	0.0109***	−0.00673	−0.0135	−0.0113	−0.0428	0.0265**	0.0266*	−0.00754
Black	0.0111***	0.00185	0.0388***	−0.00650	−0.00789	−0.00820	−0.0117	−0.0221*	0.0286	−0.0150	0.00960	−0.153***
Hispanic	−0.00892	−0.00535	−0.0267	−0.00960	−0.0110*	0.00560	0.00377	0.00408	0.0479	−0.00575	−0.0189	0.0672
Asian	0.0134***	0.0160***	0.0126	0.00711	0.00790	0.00932	0.0214	0.0292*	−0.0337	−0.00579	−0.0173	0.0934
Academy	−0.0604***	−0.0590***	−0.0581***	−0.011***	−0.0116***	0.00745	0.0263***	0.0237***	0.0825	0.0693***	0.0684***	0.0928
ROTC	−0.0140***	−0.0113***	−0.0289***	0.00360	0.00462	−0.000121	0.0194**	0.0177**	0.0336	0.00830	0.00776	0.0161
OCS/OCT	0.0140***	0.0108***	0.0293**	−0.0077**	−0.0103***	0.0104	−0.0271***	−0.0221**	−0.112*	0.0131	0.0128	−0.00977
Other source of commission	0.0604***	0.0595***	0.0577***	0.0146***	0.0173***	−0.0178	−0.0186	−0.0194	−0.00452	−0.0907***	−0.0890**	−0.0991

Table B.2—Continued

	Probability Retained as O3			Probability Promoted to O4			Probability Retained as O5			Probability Promoted to O6		
Variable	Pooled	Males	Females	Pooled	Males	Females	Pooled	Males	Females	Pooled	Males	Females
Army	−0.0325***	−0.0363***	−0.0120	0.0579***	0.0574***	0.0558***	−0.0284***	−0.0284***	−0.0301	−0.00998	−0.0112	−0.0599
Navy	−0.00757***	−0.00733***	−0.0275**	−0.065**	−0.0645***	−0.0752***	−0.0464***	−0.0463***	−0.0260	−0.0256**	−0.0206	−0.106
Marines	−0.00951**	−0.00650*	−0.0135	−0.026***	−0.0260***	−0.0213	0.0689***	0.0686***	0.0265	−0.00409	−0.00899	0.192
Air Force	0.0495***	0.0501***	0.0531***	0.0329***	0.0332***	0.0407***	0.00590	0.00608	0.0296	0.0397***	0.0408***	−0.0266
Bachelor's degree only	0.00606**	0.00786***	0.000135	−0.021***	−0.0201***	−0.0253***	−0.00395	−0.00432	0.00719	−0.0406***	−0.0392***	−0.0623**
More than bachelor's degree	−0.00606**	−0.00786***	−0.000135	0.0212***	0.0201***	0.0253***	0.00395	0.00432	−0.00719	0.0406***	0.0392***	0.0623**
Tactical occupation	0.0401***	0.0386***	0.0673***	−0.030***	−0.0273***	−0.0519***	0.0281***	0.0296***	−0.00614	0.0139	0.00128	0.0312
Intelligence occupation	−0.00746*	0.000347	−0.0380***	0.0180***	0.0205***	0.0161	−0.0248**	−0.0267**	−0.00949	0.0234	0.00839	0.0419
Engineering occupation	−0.0252***	−0.0263***	−0.0211*	0.000814	0.00134	0.00641	0.0156	0.0164	0.00215	0.0213	0.0313*	−0.0716
Administration occupation	−0.0248***	−0.0304***	−0.0145	0.00546	−0.00324	0.0336***	0.00291	7.68e−05	0.0119	0.00201	0.0250	0.0201
Supply/ procurement occupation	−0.0236***	−0.0253***	−0.0128	0.00587	0.00595	0.0158	−0.0124	−0.0136	−0.00962	−0.0598	0.00712	−0.0450

Table B.2—Continued

	Probability Retained as O3			Probability Promoted to O4			Probability Retained as O5			Probability Promoted to O6		
Variable	Pooled	Males	Females	Pooled	Males	Females	Pooled	Males	Females	Pooled	Males	Females
Other DoD occupation	0.0409***	0.0430***	0.0192	6.54e−05	0.00278	−0.0201	−0.00948	−0.00572	0.0112	−0.0590	−0.0731	0.0234
Entry year 1	0.111***	0.0976***	0.218***	−0.141***	−0.145***	−0.110***	0.0797***	0.0749***	0.162***	−0.216***	−0.224***	−0.0851
Entry year 2	0.125***	0.113***	0.220***	0.130***	0.122***	0.189***	0.105***	0.0945***	0.233***	0.0601***	0.0523**	0.0836
Entry year 3	0.133***	0.125***	0.216***	0.119***	0.116***	0.139***	0.0723***	0.0716***	0.0856	−0.0463**	−0.0519**	−0.0182
Entry year 4	0.0925***	0.0856***	0.151***	0.119***	0.117***	0.127***	0.0124	0.0127	0.00250	0.0229	0.0332	−0.0592
Entry year 5	0.0648***	0.0623***	0.0788***	0.138***	0.135***	0.166***	−0.0603***	−0.0544***	−0.105*	0.0711***	0.0677***	0.0834
Entry year 6	0.0145***	0.0126**	0.0300	0.176***	0.175***	0.195***	−0.0252*	−0.0292**	0.00813	0.0396**	0.0560***	−0.0844
Entry year 7	−0.0406***	−0.0402***	−0.0532***	0.172***	0.175***	0.145***	−0.0919***	−0.0860***	−0.172***	−0.00146	0.00256	−0.0421
Entry year 8	−0.111***	−0.105***	−0.159***	0.155***	0.157***	0.149***	−0.114***	−0.105***	−0.212***	0.0382**	0.0372**	0.0278
Entry year 9	−0.109***	−0.101***	−0.183***	0.157***	0.158***	0.139***	−0.0774***	−0.0718***	−0.153***	0.00870	0.0135	−0.0495
Entry year 10	−0.108***	−0.101***	−0.168***	0.130***	0.133***	0.104***	0.0324**	0.0238	0.106	−0.0110	−0.0101	−0.0291
Entry year 11	−0.118***	−0.105***	−0.207***	0.121***	0.127***	0.0718***	0.0669	0.0687	0.0444	0.0345	0.0233	0.173
Entry year 12	−0.0727***	−0.0596***	−0.154***	0.0741***	0.0828***	0.0148						
Entry year 13	0.124***	0.114***	0.138***	−0.718***	−0.725***	−0.697***						
Entry year 14	−0.106***	−0.0993***	−0.129***	−0.632***	−0.629***	−0.632***						

Table B.2—Continued

Variable	Probability Retained as O3			Probability Promoted to O4			Probability Retained as O5			Probability Promoted to O6		
	Pooled	Males	Females	Pooled	Males	Females	Pooled	Males	Females	Pooled	Males	Females
Constant	0.929***	0.908***	0.975***	0.929***	0.705***	0.702***	0.850***	0.843***	0.756***	0.414***	0.395***	0.488***
Total observations	68,559	68,487	93,021	92,585	68,190	55,714	35,268	32,563	11,653	10,045		

NOTE: The table shows estimates for the categorical variables that are transformed so that the estimate shown is the deviation from the grand mean for the category. For example, the coefficient for tactical DoD occupational group is the estimate for this group relative to the overall mean effect for all occupation groups.

*** Statistically significant from the zero (in the case of noncategorical variables) and from the grand mean effect (for categorical variables) at the 1-percent level.

** Statistically significant from the zero (in the case of noncategorical variables) and from the grand mean effect (for categorical variables) at the 5-percent level.

* Statistically significant from the zero (in the case of noncategorical variables) and from the grand mean effect (for categorical variables) at the 10-percent level.

Table B.3
Blinder-Oaxaca Decomposition with Detailed Decompositions, by Selected Milestone

	Probability Retained as an O3		Probability Promoted to O4		Probability Retained as an O5		Probability Promoted to O6	
Predicted difference between male and female officers	0.11788***		0.06024***		0.10880***		0.01277***	
	Explained Difference	Unexplained Difference	Explained Difference	Unexplained Difference	Explained Difference	Unexplained Difference	Explained Difference	Unexplained Difference
Explained/ unexplained difference	0.07830***	0.039580***	0.030120***	0.030120***	0.03307***	0.075730***	0.04829***	−0.035520*

	Probability Retained as an O3		Probability Promoted to O4		Probability Retained as an O5		Probability Promoted to O6	
Variable	Explained Difference	Unexplained Difference	Explained Difference	Unexplained Difference	Explained Difference	Unexplained Difference	Explained Difference	Unexplained Difference
Cumulative months of deployment	−0.000190*	−0.012530***	0.005130***	−0.00379	0.00650**	−0.00435	0.040930***	−0.030500*
Married	0.008750***	0.047700***	0.004450***	0.019950***	0.00253	0.01805	0.012990***	0.039430**
Not married	0.008750***	−0.02413	0.004450***	−0.00942	0.00253	−0.00445	0.01299	−0.01035

Table B.3—Continued

	Probability Retained as an O3		Probability Promoted to O4		Probability Retained as an O5		Probability Promoted to O6	
Variable	Explained Difference	Unexplained Difference	Explained Difference	Unexplained Difference	Explained Difference	Unexplained Difference	Explained Difference	Unexplained Difference
Any dependents ages 0–3	0.004580***	0.00212	0.000590*	0.006000***	−0.000650**	0.007200***	0.00051	0.004810**
No dependents ages 0–3	0.004580***	−0.00797	0.000590*	−0.01883	−0.000650**	−0.10135	0.00051	−0.10077
Any dependents ages 4–6	0.004760***	−0.003410***	0.001440***	0.00003	−0.000540**	0.00188	0.000770*	0.00218
No dependents ages 4–6	0.004760***	0.02277	0.001440***	−0.00095	−0.000540**	−0.01441	0.000770*	−0.01877
Any dependents ages 7–18	0.006120***	−0.00166	0.000960***	0.00094	−0.00226	−0.00671	0.00296	0.022260**
No dependents ages 7–18	0.006120***	0.00821	0.000960***	−0.00318	−0.00226	0.00787	0.00296	−0.02584
Any dependents ages 18+	−0.000840***	−0.001420***	0.000430***	−0.002490***	0.003390***	0.00265	−0.005750***	−0.00392
No dependents ages 18+	−0.000840***	0.02227	0.000430***	0.02856	0.003390***	−0.01228	−0.005750***	0.01851
Prior service	0.001520***	−0.006310***	0.001200***	−0.00239	−0.00086	−0.004750*	0.00012	−0.00378

Table B.3—Continued

Variable	Probability Retained as an O3		Probability Promoted to O4		Probability Retained as an O5		Probability Promoted to O6	
	Explained Difference	Unexplained Difference	Explained Difference	Unexplained Difference	Explained Difference	Unexplained Difference	Explained Difference	Unexplained Difference
No prior service	0.001520***	0.03085	0.001200***	0.00931	−0.00086	0.03070	0.00012	0.02739
White	−0.001720***	0.00900	0.001170***	0.012560*	−0.00111	0.02593	0.002350*	0.02761
Black	−0.000890**	−0.00487	0.00065	0.00020	0.00082	−0.00562	0.00118	0.019640***
Hispanic	0.000010***	0.00083	0.00000	−0.00062	−0.00003	−0.00101	0.00004	−0.00189
Asian	−0.000380***	0.00028	−0.00021	−0.00018	−0.00012	0.00222	0.00002	−0.00383
Academy	−0.001060***	−0.00017	−0.000310**	−0.00400	0.002770***	−0.00963	0.007260***	−0.00417
ROTC	0.000820***	0.00929	−0.00019	0.00248	0.001380**	−0.00605	0.00051	−0.00326
OCS/OCT	0.000370***	−0.003520*	−0.00008	−0.004520*	0.005250***	0.04014	−0.00242	0.00993
Other source of commission	0.000910***	0.00005	0.000190**	0.00151	−0.00032	−0.00012	−0.001670**	0.00010
Army	0.000700***	−0.007710**	−0.001480***	0.00050	−0.001180**	0.00034	−0.00041	0.00900
Navy	−0.000300***	0.004410*	−0.002340***	0.00231	0.002780***	−0.01008	0.001560*	0.04192
Marines	−0.000580***	0.00050	−0.001720***	−0.00017	0.001770***	0.00050	−0.00011	−0.00256

Table B.3—Continued

Variable	Probability Retained as an O3		Probability Promoted to O4		Probability Retained as an O5		Probability Promoted to O6	
	Explained Difference	Unexplained Difference	Explained Difference	Unexplained Difference	Explained Difference	Unexplained Difference	Explained Difference	Unexplained Difference
Air Force	−0.003900***	−0.00128	−0.002500***	−0.00330	−0.00004	−0.00691	−0.00033	0.02054
Bachelor's degree only	0.000180**	0.00707	−0.000740***	0.00475	−0.00034	−0.00195	−0.003480***	0.00401
More than bachelor's degree	0.000180**	−0.00063	−0.000740***	−0.00046	−0.00034	0.00956	−0.003480***	−0.01914
Tactical occupation	0.012170***	−0.008330**	−0.009360***	0.007780*	0.014000***	0.00512	−0.00040	−0.00267
Intelligence occupation	0.00038	0.004220**	−0.000770***	0.00036	0.00103	−0.00169	−0.00059	−0.00312
Engineering occupation	0.001260***	−0.00098	−0.000040***	−0.00101	−0.00041	0.00225	−0.00078	0.016750*
Administration occupation	0.003640***	−0.00254	−0.00088	−0.006900**	−0.00112	−0.00412	−0.00825	0.00072
Supply/ procurement occupation	0.001690***	−0.00209	−0.00044	−0.00173	0.00055	−0.00062	−0.00009	0.00826

Table B.3—Continued

Variable	Probability Retained as an O3		Probability Promoted to O4		Probability Retained as an O5		Probability Promoted to O6	
	Explained Difference	Unexplained Difference	Explained Difference	Unexplained Difference	Explained Difference	Unexplained Difference	Explained Difference	Unexplained Difference
Other DoD occupation	0.00066	0.00038	0.00000	0.00036	0.00001	−0.00011	0.00004	−0.00057
Entry year 1	0.00011	−0.001550***	0.00019	−0.00063	0.002730***	−0.00632***	−0.005990**	−0.013070**
Entry year 2	0.00024	−0.005680***	−0.001070**	−0.004870***	0.00025	−0.00742***	−0.00036	−0.00214
Entry year 3	0.002580***	−0.004580***	0.0016800***	−0.00158	−0.001540**	−0.00097	0.00117	−0.00254
Entry year 4	0.001680***	−0.001130***	0.001210**	−0.00078	−0.00042	0.00090	−0.00083	0.00821
Entry year 5	0.001540***	−0.00126	0.002820***	−0.00240	0.001420**	0.00398	−0.001270*	−0.00109
Entry year 6	0.000230***	0.00111	0.002020***	−0.00161	0.00043	−0.00299	−0.00089	0.011830**
Entry year 7	−0.00071	0.00614	0.002330***	0.002680*	−0.00162	0.01383***	−0.00003	0.00661
Entry year 8	−0.00049	0.010580***	0.00112	0.00070	−0.00066	0.02091***	0.00094	0.00148
Entry year 9	0.00065	0.009310***	0.00043	0.00206	−0.00271***	0.01267***	0.00035	0.00917
Entry year 10	0.001940***	0.015340***	−0.001350**	0.003560*	−0.00001	−0.00349***	0.00009	0.00100
Entry year 11	0.004680***	0.001680***	−0.002450***	0.006640***	0.00006	0.00002	0.00003	−0.00019

Table B.3—Continued

Variable	Probability Retained as an O3		Probability Promoted to O4		Probability Retained as an O5		Probability Promoted to O6	
	Explained Difference	Unexplained Difference	Explained Difference	Unexplained Difference	Explained Difference	Unexplained Difference	Explained Difference	Unexplained Difference
Entry year 12	0.000680***	−0.000160**	−0.000650***	0.00121				
Entry year 13	−0.000920***	0.00085	0.007800***	−0.00031				
Entry year 14	0.002320***	−0.06717	0.012550***	0.00003				
Constant		0.039580***		−0.00822		0.08642		−0.092710***

*** Statistically significant from the zero at the 1−percent level.
** Statistically significant from the zero at the 5−percent level.
* Statistically significant from the zero at the 10−percent level.

Abbreviations

DEERS	Defense Enrollment Eligibility Reporting System
DMDC	Defense Manpower Data Center
DoD	Department of Defense
FY	fiscal year
OCS/OCT	Officer Candidate School/Officer Candidate Training
ROTC	Reserve Officer Training Corps

Bibliography

"Air Force Considering Longer Maternity Leave," *Air Force Times*, July 8, 2015. As of December 21, 2015:
http://www.airforcetimes.com/story/military/2015/07/08/air-force-considering-longer-maternity-leave/29869643/

Asch, Beth, Trey Miller, and Alessandro Malchiodi, *A New Look at Gender and Minority Differences in Officer Career Progression in the Military*, Santa Monica, Calif.: RAND Corporation, TR-1159-OSD, 2012. As of July 28, 2015:
http://www.rand.org/pubs/technical_reports/TR1159.html

Blinder, Alan, "Wage Discrimination: Reduced Form and Structural Estimates," *Journal of Human Resources*, Vol. 8, 1973, pp. 436–455.

Buckles, Kasey, "Understanding the Returns to Delayed Childbearing for Working Women," *American Economic Review*, Vol. 98, No. 2, 2008, pp. 403–407.

Cotton, Jeremiah, "On the Decomposition of Wage Differentials," *Review of Economics and Statistics*, Vol. 70, No. 2, 1988, pp. 236–243.

Department of Defense, *Population Representation in the Military Services: Fiscal Year 2013 Report*, Washington, D.C., 2015.

DoD—*See* Department of Defense.

Dugan, Beverly A., Jessica L. Terner, Felicity A. Tagliarini, J. Anthony Bayless, and Helene R. Felber, *The Glass Ceiling: Potential Causes and Possible Solutions*, Alexandria, Va.: United States Army Research Institute for the Behavioral and Social Sciences, AD-A278 051, 1993.

Fortin, Nicole, "The Gender Wage Gap Among Young Adults in the United States: The Importance of Money Versus People," *Journal of Human Resources*, Vol. 43, No. 4, 2007, pp. 884–102.

Fortin, Nicole, Thomas Lemieux, and Sergio Firpo, "Decomposition Methods in Economics," in Orley Ashenfelter and David Card, eds., *Handbook of Labor Economics*, Vol. 4, Part A, Philadelphia: Elsevier B.V., 2011, pp. 1–102.

Goehring, Dwight J., *Measuring Institutional Discrimination in the Army: 1974–1980,* Alexandria, Va.: United States Army Research Institute for the Behavioral and Social Sciences, AD-A151 042, 1981.

Golding, Heidi L. W., James L. Gasch, David Gregory, Anita U. Hattiangadi, Thomas A. Husted, Carol S. Moore, Robert W. Shuford, and Daniel A. Seiver, *Fleet Attrition: What Causes It and What to Do About It*, Alexandria, Va.: Center for Naval Analyses, CRMD0004216.A2, 2001.

Hattiangadi, Anita U., Gary Lee, and Aline O. Quester, *Recruiting Hispanics: The Marine Corps Experience Final Report*, Alexandria, Va.: Center for Naval Analyses, CRM D0009071.A2, 2004.

Heckman, James, "Shadow Prices, Market Wages, and Labor Supply," *Econometrica*, Vol. 42, 1974, pp. 679–694.

———, "The Common Structure of Statistical Models of Truncation, Sample Selection, and Limited Dependent Variables and a Simple Estimator for Such Models," *Annals of Economic and Social Measurement*, Vol. 5, 1976, pp. 475–492.

Horace, William, and Ronald Oaxaca, "Results on the Bias and Inconsistency of Ordinary Least Squares for the Linear Probability Model," *Economics Letters*, Vol. 90, No. 3, 2006, pp. 321–327.

Hosek, Susan D., Peter Tiemeyer, M. Rebecca Kilburn, Debra A. Strong, Selika Ducksworth, and Reginald Ray, *Minority and Gender Differences in Officer Career Progression*, Santa Monica, Calif.: RAND Corporation, MR-1184-OSD, 2001. As of September 30, 2015:
http://www.rand.org/pubs/monograph_reports/MR1184.html

Jann, Ben, "DEVCON: Stata Module to Apply the Deviation Contrast Transform to Estimation Results," unpublished Stata ado file, 2005. As of December 21, 2015:
https://ideas.repec.org/c/boc/bocode/s450603.html#cites

———, "The Oaxaca-Blinder Decomposition for Linear Regression Models," *Stata Journal*, Vol. 8, 2008, pp. 435–479.

Jones, June Taylor, *Gender Differences in Job Satisfaction in the U.S. Army*, Alexandria, Va.: United States Army Research Institute for the Behavioral and Social Sciences, AD-A339 232, 1997.

Kraus, Amanda, Ann D. Parcell, David L. Reese, and Robert W. Shuford, *Navy Officer Diversity and the Retention of Women and Minorities: A Look at the Surface Warfare and Aviation Communities*, Alexandria, Va.: Center for Naval Analysis, DRM-2013-U-005306, 2013.

Lim, Nelson, Louis T. Mariano, Amy G. Cox, David Schulker, and Lawrence M. Hanser, *Improving Demographic Diversity in the U.S. Air Force Officer Corps*, Santa Monica, Calif.: RAND Corporation, RR-495-AF, 2014. As of September 17, 2015:
http://www.rand.org/pubs/research_reports/RR495.html

Long, Valarie A., "Retention and the Dual-Military Couple: Implications for Military Readiness," in James E. Parco and David A. Levy, eds., *Attitudes Aren't Free: Thinking Deeply About Diversity in the US Armed Forces*, Maxwell Air Force Base, Ala.: Air University Press, 2010, pp. 349–362.

Losey, Stephen, "Career Intermission Applications Start July 1," *Air Force Times*, June 16, 2015. As of January 26, 2016:
http://www.airforcetimes.com/story/military/2015/06/16/career-intermission-applications-start-july-1/28801583/

Lundberg, Shelly, and Elaina Rose, "Parenthood and the Earnings of Married Men and Women," *Labour Economics*, Vol. 7, 2000, pp. 689–710.

Marshall-Mies, Joanne C., Jennifer A. Martin, and Tanya B. Seligson, *Racial Differences in Job Satisfaction*, Alexandria, Va.: United States Army Research Institute for the Behavioral and Social Sciences, AD-A383 184, 2000.

Military Leadership Diversity Commission, *From Representation to Inclusion: Diversity Leadership for the 21st Century Military, Final Report*, Arlington, Va., 2011.

Miller, Amalia, "The Effects of Motherhood Timing on Career Path," *Journal of Population Economics*, Vol. 24, No. 3, 2011, pp. 1071–1100.

Miller, Laura L., Sarah O. Meadows, Lawrence M. Hanser, and Stephanie L. Taylor, *Year of the Air Force Family: 2009 Survey of Active-Duty Spouses*, Santa Monica, Calif.: RAND Corporation, TR-879-AF, 2011. As of December 21, 2015:
http://www.rand.org/pubs/technical_reports/TR879.html

Moini, Joy S., Gail L. Zellman, and Susan M. Gates, *Providing Child Care to Military Families The Role of the Demand Formula in Defining Need and Informing Policy*, Santa Monica, Calif.: RAND Corporation, MG-387-OSD, 2006. As of December 21, 2015:
http://www.rand.org/pubs/monographs/MG387.html

Navy Personnel Command, "21st Century Sailor," web page, n.d. As of January 26, 2016:
http://www.public.navy.mil/bupers-npc/support/21st_Century_Sailor/Pages/default.aspx

Neuman, Shoshana, and Ronald Oaxaca, "Wage Decompositions with Selectivity-Corrected Wage Equations: A Methodological Note," *Journal of Economic Inequality*, Vol. 2, 2004, pp. 3–10.

Neumark, David, "Employer's Discriminatory Behavior and the Estimation of Wage Discrimination," *Journal of Human Resources*, Vol. 23, No. 3, 1988, pp. 279–295.

Oaxaca, Ronald, "Male-Female Wage Differentials in the Urban Labor Market," *International Economic Review*, Vol. 14, 1973, pp. 693–709.

Oaxaca, Ronald, and Michael Ransom, "On Discrimination and the Decomposition of Wage Differentials," *Journal of Econometrics*, Vol. 61, No. 1, 1994, pp. 5–21.

Office of the Under Secretary of Defense, Personnel and Readiness, *Report to Congress on the Review of Laws, Policies and Regulations Restricting the Service of Female Members in the U.S. Armed Forces*, Washington, D.C.: Department of Defense, February 2012.

Quester, Aline O., James H. North, and Theresa H. Kimble, *Identifying Successful Marine Corps Recruits*, Alexandria, Va.: Center for Naval Analyses, CRM 89-314, 1990.

Rachford, Douglas L., *Gender Analysis of the Professional Development of Officers Study (PDOS) Survey*, Alexandria, Va.: United States Army Research Institute for the Behavioral and Social Sciences, AD-A168 878, 1985.

Smith, David G., *Developing Pathways to Serving Together: Dual Military Couples' Life Course and Decision-Making*, dissertation, College Park: University of Maryland, College Park, 2010.

Steinberg, Alma G., Beverly C. Harris, and Jacquelyn Scarville, *Why Promotable Female Officers Leave the Army*, Alexandria, Va.: United States Army Research Institute for the Behavioral and Social Sciences, AD-A268 946, 1993.

Teplitzky, Martha L., *Dual Army Career Couples: Factors Related to the Career Intentions of Men and Women*, Alexandria, Va.: United States Army Research Institute for the Behavioral and Social Sciences, AD-A199 289, 1988.

Teplitzky, Martha L., Shelley A. Thomas, and Glenda Y. Nogami, *Dual Army Career Officers: Job Attitudes and Career Intentions of Male and Female Officers*, Alexandria, Va.: United States Army Research Institute for the Behavioral and Social Sciences, AD-A199 071, 1988.

Waite, Linda, "Does Marriage Matter?" *Demography*, Vol. 32, No. 4, 1995, pp. 483–507.

Wild, Elizabeth Ty, David T. Ellwood, and Lily Batchelder, *The Mommy Track Divides: The Impact of Childbearing on Wages of Women of Differing Skill Levels*, Working Paper 16582, Cambridge, Mass.: National Bureau of Economic Research, December 2010.

Yun, Myeong-Su, "A Simple Solution to the Identification Problem in Detailed Wage Decomposition," *Economic Inquiry*, Vol. 43, 2005, pp. 766–772.

Broadcasting Through Crisis

Broadcasting Through Crisis

How to Keep Going When Tragedy Hits

Ann S. Utterback, Ph.D.

Bonus Books

Los Angeles, California

This book has been researched and written with the best medical and scientific information and assistance available to the author. However, this book offers recommendations only, and not substitutions for conventional medical guidance. Do not use this book to diagnose or treat a physical or mental health problem. All matters regarding your health require medical supervision; you should always consult a physician when making any changes in your diet, exercise routine, or treatment of illness. This book is sold with the understanding that the author is not engaged in rendering psychological, medical, or health services. If expert assistance or counseling is needed, the reader should seek assistance from a professional in the appropriate field. No responsibility will be taken by the author.

Credits

"What Will My World Be Today," copyright 2003, Maxbox Music BMI, written by Fred Bogert, 242 W. Main St., PMB 242, Hendersonville, TN 37075.

Cartoon on page 68 by permission of Dave Coverly and Creators Syndicate Inc.

"Covering a Terrorist Incident," copyright 2003, NewsLab. Reprinted with permission of the executive director of NewsLab, Deborah Potter. You can find out more about NewsLab at *www.newslab.org*.

Library of Congress Control Number: 2005925571

ISBN: 1-56625-240-7

09 08 07 06 05 5 4 3 2 1

Bonus Books
1223 Wilshire Blvd. #597
Santa Monica, CA 90403

Printed in the United States of America

Table of Contents

On the Front Lines When Severe Weather Strikes

Lessons from the Wars

CRISIS SURVIVAL KIT

Foreword

by Jim Farley

On the morning of September 11, 2001, the first pictures of smoke coming from atop the World Trade Center filled all the TV monitors in the WTOP Radio newsroom in Washington, D.C. Staffers from other departments at the radio station poured into the newsroom and stood, transfixed, as the second plane hit. Not at all transfixed were the regular denizens of the newsroom: reporters, editors, anchors, and writers. It was as if the news folks answered a fire bell and sprang into action. They were rushing around their immobile, stunned brethren from other departments, running copy into the main air studio and bringing the activity level up to a controlled frenzy.

Reporters raced to the Pentagon, the White House, Capitol Hill, and other locations that might also have been hit. Staffers who were off for the day or on vacation began to show up without being asked and pitched right in. In short, it's what one would hope for from a well-trained staff of both experienced and young journalists.

By midday some people had been working nonstop since midnight, and I had a hard time getting them to go home. "This is not a sprint. It's a marathon," I told them. "We have to have people who are fresh to do this again tomorrow morning." Those who did go home did not sleep. They remained glued to the TV and radio. For the weeks that followed, they ran on coffee, donuts, pizza, caffeinated soda, and adrenaline.

The differences between the news staffers and the non-news staffers were still apparent beyond the first day. Within a few days, the non-news staffers were able to grieve. They cried over the senseless deaths of so many people at the World Trade Center, the Pentagon, and the field in Pennsylvania. On the other hand, many of my news staffers would later tell me it had taken them weeks before they could cry. Their grieving was delayed because they were involved with the story nonstop. There was no time for personal reflection.

In a few months Washington was gripped with a new fear as people sickened and died from anthrax found on Capitol Hill and in the Brentwood Post Office. Plastic sheeting and duct tape were the topics as Homeland Security officials gravely told us Washington was a likely terrorist target, and mobile antiaircraft missile batteries sprouted on the National Mall.

Just months after that, in October 2002, the entire Washington region was plunged into fear by the sniper shootings. In some respects this was worse than the terrorist attacks of September 11, because the target was anyone and everyone going about his or her normal activities. It was random, and the chilling message from the snipers was "Your children are not safe, anywhere at any time." By now I could really see the stress and strain on the reporters, who went out and covered people who were being told to zigzag as they crossed parking lots, and duck behind the gas pumps as they filled their cars because that's where the snipers had struck. My staff members were assuring their families at home they were fine because they did not want to worry them. But they were as stressed as anybody else in the region. They just felt like they could not express it out loud.

In a radio or TV newsroom there are on-air journalists and off-air journalists. The stress was not limited to those on the air. I have always told off-air editors and producers, "You see that journalist over there, that on-air anchor [or reporter]? He or she is your coequal, fellow journalist until the on-air light goes on. At that point he or she is an on-air performer as well. Those of us who work off air have got to keep them calm and confident."

Ann Utterback knows this, and she knows how all journalists can cope with stress and strain to maintain and extend their professional lives. Shortly after September 11, in November 2001, we had her come to WTOP to give a seminar for anchors and reporters entitled "Broadcasting Through a Crisis: Calm Voices in Tense Times." I had heard Ann speak at the Chesapeake Associated Broadcasters Convention a year earlier, and based on her advice about keeping voices healthy I immediately ordered a bottled water dispenser for the main air studio. When she did her workshop on "Broadcasting Through a Crisis," it was a big hit with our stressed staffers (read, all of them). Ann brought a massage therapist with her who worked on staffers during her talk to help them relax, which was also beneficial.

We've loaned every new staff member a copy of Ann's previous book *Broadcaster's Survival Guide,* and we'll no doubt order a dozen copies of this book. You can't stay in this business if you can't stand the stress. Ann knows how to help you cope with that stress and to succeed and grow professionally. And now she also tells us how those who went before us in this business not only coped but also succeeded beyond all expectations.

JIM FARLEY
VICE PRESIDENT OF NEWS AND PROGRAMMING
WTOP RADIO, WASHINGTON, D.C.

Preface

Writing this book has been a very fulfilling experience for me. With crisis events happening with intensity in the last few years, I wanted to be able to give something to the community of broadcasters with whom I have worked for two decades. I think that witnessing crises makes all of us want to help in some way. This is my way to help.

This book covers more areas than any one person could be an expert in. I am not a fitness or nutrition professional, for example, but I've had the good fortune to work with many of them. I also have an unquenchable desire to learn about these areas. What I've done in this book is pass on all the knowledge I could gather on many different subjects.

Nothing in this book will be shockingly new to you. Most of what I've included is fairly common knowledge. But, if you're like most people, you are too busy to read about all these areas in a diverse collection of sources. You may also be too busy to think of applying everything you read. My goal is to present this information in a way that's easy to understand and simple to apply. During a crisis situation, the recommendations you'll read here can make the crucial difference in your on-air performance and your health.

The best time to apply the tips you'll read is now. Begin by making one change in your lifestyle a week. The changes you choose to make will be the right ones for you. Lifestyle changes don't follow the pattern of "one size fits all." Start with small changes that you think will work for you and build up to larger ones. You should always consult your physician when making any big changes in diet, exercise, or lifestyle. This book is a supplement, not an alternative, to your regular medical care.

Most small changes you'll make are simple, but they may not be easy. It's much easier to keep living life in the same comfortable fast lane that you're probably used to by now. As I say many times in this book, you have to put yourself in a training mode and stay in training to take the best care of yourself in this stressful career. If you were in a less stressful career, for example, it might seem like a luxury to take time for yourself to relax. In this business, it's not a luxury. It's a necessity.

I hope you find this book helpful and that it provides a life preserver in the turbulent times in which we live. I guarantee you that the recommendations you'll read can make a difference in your life. They've made a difference in mine.

Acknowledgments

Crises have been a part of the news business since the beginning. Certainly in my life I've watched the unfolding of what appears to be the worst crisis that can happen, followed quickly by another that tops it. In 1963, for example, we thought nothing could be worse than the assassination of a favorite president, but the events of 1968 topped that. I was only a few miles away when Dr. Martin Luther King was killed in Memphis. I remember climbing out on the roof to watch the fires from riots that followed. It seemed the world was coming to an end.

A broadcasting professor of mine brought a new perspective into my life during that crisis. David Yellin explained to me that this was a time to remain a calm observer of this momentous event. He recruited many of his students to collect every scrap of material we could find that related to the assassination. We combed the streets picking up picket signs the garbage workers had carried when Dr. King came to help them and any scrap that might be significant. We audiotaped interviews, and we took photographs. All this history is now stored in the Mississippi Valley Collection at the University of Memphis library and has provided a priceless record of that time. David Yellin showed me how to be a reliable recorder of events during a crisis. I will always be grateful to him and his wife, Carol Lynn.

I also have many people to thank who helped with this book both directly and indirectly. First, I want to thank all of my clients over the years, who have really been my teachers. They have shown me what it's like to work as a broadcaster dealing with the stresses that accompany the job.

The contributors to this book are many, but I want to thank them all. Mike Freedman has provided a complete view of war coverage from World War II to the present, drawing on some of the luminaries in the field of war reporting. Two broadcast managers who have literally "weathered" the storms of broadcasting, Elliott Wiser and Dan Shelley, have written very helpful chapters on covering severe weather. Deborah Potter was kind enough to allow me to reprint the NewsLab article on covering terrorist incidents. I consider Jim Farley to be one of the best news managers in the business, and I am honored that he wrote the foreword. And my "Reporters in the Trenches" have provided first-hand accounts of what it's like to cover crisis events. I wish to say a special thank-you to them: Mike Walter, Melinda Murphy, Sherry Stalley, Neal Augenstein, Ross Simpson, and Desirée Berenguer. Their profiles touched me deeply, and I hope they will affect you as well. My thanks also to Fred Bogert for his wonderful music and the rights to print the lyrics from his inspiring song, "What Will My World Be Today."

Two people are especially important to me personally. Elizabeth and J have helped me grow into all that I am, and I am grateful for that every day.

This is the eighth book I've written, and the one person who always receives my deepest appreciation is my husband Jim. He is my best editor and my best friend of over forty years. He's my comfort in a storm.

Ann S. Utterback
Washington, D.C.

"What is to give light, must endure burning."

Victor Frankl

*"The world breaks around everyone,
and afterward many are strong at the broken places."*

Ernest Hemingway

CHAPTER

1

An Occupational Hazard: Crisis Stress

"Anybody who covered 9/11 took a field trip through hell."

Documentary *The Day the Towers Fell*

Crisis is a given in the field of broadcasting. In many ways, it's what makes the profession so exciting. Some people are drawn to the field because of this excitement. But any crisis brings with it stress. Every profession has its occupational hazards, and for broadcasters, stress is the winner. The past few years have produced some of the biggest crises our country has seen, and along with them some of the most devastating experiences for broadcasters.

As the entire world gazed in horror at the events of September 11, along with concerns about the victims, my thoughts turned to my clients, who would surely face months of some of the toughest coverage they'd ever have to do. They would be working long hours that allowed little time for themselves and their own anguish. They would have to face work overload and exposure to unimaginable human suffering.

Weeks later, watching tributes to the emergency workers from September 11 also made me think of my clients. At those tributes no one was empathizing with the journalists' struggles, even though many of them had been elbow to elbow with the emergency workers. No one was considering how this might affect them.

And no one was considering the fact that, in broadcast news, one crisis follows the next. Yes, September 11 was a huge crisis, but look what followed it: anthrax, the threat of more terrorism, the D.C. sniper, the Iraq war, SARS, the scandal at Abu Ghraib prison, and many other big stories. The stress in this business does not let up. That's something most young broadcasters have to learn. Veteran broadcaster Bob Schieffer talks about this in his autobiography, *This Just In.* He says that in 1962, when he covered the riots at the University of Mississippi as James Meredith was being enrolled as its first African-American student, he felt that "we were covering the story of our lives." But in a year he was covering the assassination of John F.

Kennedy, and many other crises followed that. Consider, for example, the subtitle of David Brinkley's 1995 memoir: *11 Presidents, 4 Wars, 22 Political Conventions, 1 Moon Landing, 3 Assassinations, 2,000 Weeks of News . . .*

And crises don't have to be the top story to have an emotional effect on you. Routine stories like fatal traffic accidents, shootings, domestic and sexual abuse, and fires all hold the potential to become stories that haunt you with their graphic details.

It never stops in broadcast news. As novelist Michael Connelly says of police work, "every case is a battle in a war that never ends." Crises present circumstances that you can't change. The hurricane will hit; wars will happen; terrorists do exist in the world. What you can change is the way that you deal with these events.

In order to continue to cover these crises both big and small, you have to realize that this business is *a marathon and not a sprint.* Even in a major catastrophe, you have to take care of yourself so that you can avoid burnout. I often tell clients that they have to *stay in training* to do this job well. Just like an athlete, you have to think about all the ways that you can stay healthy and emotionally balanced.

NO FREAKING OUT

(Artwork courtesy of Alaska Public Radio Network.)

Staying in training is not just for your benefit, either. You have a responsibility to remain focused on the most important part of your job, which is to convey information to the listener in a calm and focused manner. You should never add to a crisis in the way you report it. Many times your tone of voice, the rate of your delivery, the volume of your voice, and your nonverbal communication convey more than the words you're saying. In a catastrophe the last thing viewers need is for broadcasters to create panic because they are portraying panic. I call it "Chichen Little Syndrome"—when an anchor or reporter makes us think the sky is falling because of his or her panicked delivery.

Television news stories have an archival nature that many broadcasters don't consider. My guess is that most families have a story or two on tape that they have saved to show to friends and family. Maybe it was a happy story about a relative winning the lottery, but it might also be a sad story about a battle in Iraq that took the life of a family member. No doubt many children will watch coverage of September 11 when they grow up to learn what happened to their mother or father. Your responsibility is to deliver these stories in the most focused and professional way that you can.

It doesn't take a tragedy of a huge proportion to call for a delivery that is focused and calm in the midst of crisis. No story is insignificant. For someone, it is the most important story he'll ever hear. You should never be shouting in your live shot or rushing your track. Imagine someone listening to this delivery every time he watches the tape. You owe him the most focused and professional job you can do.

One of the reasons that we fail to respond well to major stresses is because we are not taking care of stress as it happens. We need to deal with stress on a daily and almost an hourly basis. One D.C. client told me, after covering September 11, the anthrax attacks, and the D.C. sniper case, that she felt like she was carrying around a garbage dump of emotions that she had not had time to deal with. She was feeling overwhelmed because she had not taken time to process all that she had been through. You can't be ready for the next crisis if you haven't dealt with the past one.

So how do you deal with the stress of this business and remain focused? I think you have to practice what life coach Thomas Leonard calls "extreme self-care." And this has to be a practice that becomes part of your daily life. Think of it this way: You wouldn't start hurricane-proofing your house once the hurricane hit, would you? So you can't start stress-proofing yourself when the stress hits. It's a process that needs to happen all the time.

The following chapters of this book will give you a way to take care of yourself both before and during a crisis. You'll learn how important relaxation, nutrition, exercise, sleep, and having fun are, both personally and professionally. And you'll learn how to deal with the feelings that are a natural part of going through any crisis. You'll find that some ways of coping with stress work better in some situations, while others are perfect at another time. If you use the methods described in this book, you'll have lots of choices. And, much like eating at a buffet, you may find that you get the most balanced meal by eating a little of everything offered.

My hope is that you'll have a roadmap to keep yourself healthy and balanced, so that, no matter what the crisis, you can withstand it. You can stay focused on your job as a broadcaster. You can go into the war zones and come out in one piece both physically and emotionally. To borrow the slogan of WETA, the D.C. PBS station, you will be able to offer "in an uncertain world, a steady voice."

CHAPTER

2

Coping with Crisis Stress

"Fear isn't rational. It's wired into our brains for survival. You can't change it with reason."

George Gray, Harvard School of Public Health

When clients began calling me and coming to my office in the first days after September 11, 2001, they were all looking for the same thing. They wanted some relief from the stress of covering that catastrophic event. They were finding that they could not perform their jobs well because of the horrors they were seeing, the long hours they were working, and the stress that was building inside them. A photojournalist speaking in the documentary *The Day the Towers Fell* expressed the level of stress I saw in my clients when he said, "Anybody who covered 9/11 took a field trip through hell." This characterization is certainly reflected in the profiles included in this book (see "Profiles of Crisis: Reporters in the Trenches, "chapters 9–11). I subsequently began doing workshops at networks to help broadcasters cope with the stress.

When I was first asked to come in and do a workshop on broadcasting through a crisis, I thought about what clients had been seeking. They wanted "relief." I took the word RELIEF and used it as an acronym for the main elements that alleviate stress. The chapters that follow will give you information about each one.

Broadcasting Through Crisis: How Do You Spell "RELIEF"?

Relax

Exercise

Let Yourself Sleep

Identify Your Feelings

Eat Well

Focus on Fun and Normalcy

Veteran reporters like Wolf Blitzer know the value of using these measures to cope with stress. During the Iraqi war in 2003, Mike Moss, a WTOP radio anchor, asked Wolf Blitzer how he was doing in Kuwait City. Wolf's response was "I'm pacing myself. I'm drinking a lot of water, exercising, and getting enough sleep. I'm doing okay." He knows that this is what it takes to stay healthy and perform well during a crisis.

Crisis stress can hit at any time. It doesn't have to be precipitated by an event on the scale of a terrorist attack or a war. You may be in a small market covering a devastating domestic dispute. You may find yourself at a gruesome traffic, train, or airplane accident (see chapter 14, "Stories That Haunt You"). Or you may simply see feed footage that traumatizes you. One client of mine does not even cover hard news. She works for a weather network, but she happened to witness the beheading of Nicholas Berg in Iraq when she looked up at the Reuters feed. And none of us is safe from nature. A natural disaster may strike in your town that leaves death and devastation everywhere. Or your personal safety may be threatened by a stalker (see Crisis Survival Kit 1). These events happen every day for broadcasters, and they can produce crisis stress.

In this chapter you'll learn that your body doesn't discriminate between stressors. It doesn't have to be the biggest trauma to produce the biggest response. Your particular genetic makeup and personal history may make you less stress-resilient, so that what one reporter might walk away from and forget may get stuck in your memory on replay, resulting in Post-Traumatic Stress Disorder (PTSD). Stress response is a very personal thing, but basic RELIEF measures can help.

Before we look at how to find RELIEF, let's explore what stress is and how it affects the body. This will provide an understanding of the problem that will make the solutions more meaningful.

WHAT IS STRESS?

One of the most beneficial ways of protecting ourselves as human beings is our *fight or flight response*—but it can also be one of the most physically destructive. The fight or flight response is a result of the hundreds of thousands of years humans spent living in dangerous places like deserts and rainforests. We needed a protective system back then that could go into action instantly. That's what our fight or flight response does. It mobilizes our bodies and turns us into fighting or running machines.

When something threatens us, the primitive part of our brain starts a cascade of physiological changes before our rational brain can make us aware something is wrong. This was a lifesaving procedure in the early years of human existence, when danger lurked behind every tree. It probably is responsible for our survival on the planet. But, now, it's too powerful for most of the threats that come our way, and scientists have found that it's hard to shut off. We now live in world that is usually *mentally,* rather than physically, stressful, and we get high-level chemicals released in our bodies for what is often a low-level threat—or no threat at all.

HOW THE BODY REACTS

Dr. Walter B. Cannon, a professor of psychology at Harvard Medical School in the early 1900s, was the first to write about the fight or flight response. He described the ways that the body responds when it is forced to adjust to physical stresses such as extreme heat or cold, great exertion, or psychological stress like when the emotions are triggered by a perceived threat.

The body has a complicated and sophisticated reaction to stress. The hypothalamus, at the base of the brain, signals the adrenal glands above the kidneys to release more than fifty stress hormones and steroids. The most important of these are adrenaline (also called epinephrine), noradrenaline, and the steroids that make up the glucocordicoids. Adrenaline and noradrenaline are the emergency hormones that help us through a crisis by being released instantly, and the glucocordicoids come into play five or ten minutes after the initial release of adrenaline.

Let's look at what this bodily response did for our early ancestors who lived in the wild. If a bear attacked one of our ancestors, he would obviously want to fight or flee. In order to do this, his body would secrete the stress hormones to propel him to action. When danger approached, his bodily systems would react instantly and simultaneously. His metabolism would increase and this would convert food into energy faster. His blood sugar (glucose level) would spike upwards. His heart would begin beating faster and harder. At the same time, the blood vessels in his extremities would constrict, allowing more blood to go to his vital organs. This would also reduce bleeding if he were injured. His blood chemistry would change so that it would clot more easily in case of injury. His breathing would speed up to allow more oxygen for quick thinking and to assist in fighting or fleeing. His digestion would shut down to divert more blood to vital functions, because as stress expert and author Robert M. Sapolsky says, "You have better things to do than digest breakfast when you're trying to avoid being someone's lunch" (see Suggested Readings). In fact all bodily systems that were not needed for survival would be put on hold. Muscles would contract to create armor for the body. And his pupils would dilate for better vision. His body would very rapidly become a fighting machine.

Now let's move that reaction into modern time. Neal Augenstein faced long hours every day staked out waiting for press conferences about the D.C. sniper (see chapter 12). Suppose he had seen a dark car circling the parking lot. His eyes observe this and signal his hypothalamus at the base of his brain to alert his body that something may be terribly wrong. He may be in danger. The hypothalamus sends the message to the adrenal glands, and they begin secreting adrenaline and the other stress hormones. Before he knows it, he's breathing faster and his hands have become sweaty. His blood pressure and heart rate have gone up and the blood vessels to his extremities have constricted. He begins to feel queasy as his digestive system shuts down. His mouth gets dry as the blood supply to the salivary glands is diverted. All his blood is being channeled to his vital organs so he can fight or flee.

Suppose after thirty seconds Neal had seen a reporter he knows emerge from the mysterious car. He would have seen there was no threat at all; the stress reaction his body had mobilized was not an appropriate response to this situation. But the fear of the sniper caused his body to go into a major survival mode instantly, before he could think through the situation. There would be no chance to recover from the physiological reaction of the fight or flight response before he had to continue with his day, facing even more stresses.

THE CHEMISTRY OF STRESS

One of the ways the fight or flight response becomes harmful to the body is in the makeup of the chemicals that are excreted. The first two, adrenaline and noradrenaline, are like the match you use to light a gas stove. They supply a burst of energy within seconds. But the glucocordicoids that follow this initial burst are more like the gas that fuels the stove. They can last a long time, and they can keep our bodies in the stress response. And it takes them hours to leave our bodies even when the stress has passed. In fact, a ten-minute-long stressor may keep glucocordicoids coursing through our systems for two hours. The long-term effects of stress make our bodies react much like the stove that is left to burn for days. Eventually the stove itself will suffer from the heat of the constant flame.

Esther Sternberg is one of the leading experts on stress, and author of a book on the subject (see Suggested Readings). She is the director of the integrative neural-immune program and chief of the neuroendocrine immunology and behavior section at the National Institutes of Health (NIH). She describes the effects of the stress hormones in the body as an inverted U-shaped curve. On the rising arm stress is initiated and performance of the body is enhanced, as we saw with our primitive man. It reaches a peak, however, and then performance in all aspects begins to decline. The first burst of adrenaline is good, but the glucocorticoids that follow erode many of the good effects. This results in the negative effects of chronic stress.

Another problem with the glucocordicoids is that we don't feel their effects the same way we do the adrenaline rush that we're all familiar with. You might know this rush from a pleasant occurrence like getting the burst of energy you need to sprint to the finish in a race or the thrill of riding a roller coaster. You might also feel it during live shots. We actually feel something similar when we consume caffeine or sugar.

A good example of what happens to our heart rate during an adrenaline rush came from a friend of mine who is an avid bicyclist. He goes on regular "century" rides of one hundred miles in a day. He wears a pulse-rate monitor when he rides, and he said he was amazed recently when a barking dog came running up to his bike. He said his heart rate is usually around 146 beats per minute when he rides. He happened to look down at the monitor when the dog approached and saw that his heart had jumped almost instantly to 166 even though he was not peddling any faster. He began to feel all the manifestations of the fight or flight response and saw it clearly on the pulse rate monitor.

For some people the adrenaline rush becomes an addiction. They may set themselves up to feel it by pushing deadlines, or keeping their police scanners running even at night when they're off work. We may like that feeling of getting the rush because adrenaline makes us less conscious of pain, both physical and emotional, and it makes life seem more vivid. Some people use this rush for fuel instead of using good fuel sources for the body like healthy nutrition, water, exercise, and relaxation.

They may not realize that the glucocorticoids that follow the adrenaline rush cruise through their bodies slowly and silently for a long time and can have detrimental effects. Stress hormones may feel good at first, but, as author Martha Beck points out, "just like heroin or aerosol fumes, stress hormones have side effects that can kill you" (see Suggested Readings).

There's a parable about a frog that illustrates the negative effects of glucocorticoids. Legend has it that if you put a frog in a pan of boiling water he will jump out. You can't trap him that way. But if you put a frog in a pan of cold water and slowly increase the temperature, the frog will stay in the water and boil. Chronic stress affects us in a similar manner. We stop registering the rush of adrenaline, so we don't realize the threat.

NEGATIVE EFFECTS OF CHRONIC STRESS

The irony of the fight or flight response is that it's so highly effective in the short term and so potentially destructive in the long term. The effect it has on the immune system is a perfect example. In the short term adrenaline heightens the efficiency of the immune system, giving us increased immunity, but the glucocorticoids change that. They actually *decrease* our immunity over time. Because of this, we're five times more likely to catch a cold or the flu when we're stressed. Most of us have experienced this many times in our lives. It's our immune systems that keep us well, not a lack of viruses or bacteria around us. In fact, we have around thirty-two million bacteria living on each square inch of our bodies. Some are good and some bad. We pick up viruses and bacteria constantly on our hands, and we could get sick at any time. When chronic stress has compromised our immune systems, we have nothing to protect us from the bacteria and viruses with which we share the planet.

Numerous scientific studies have proven that stress causes a weakening of the immune system. One such study at Ohio State University found that the body actually heals slower during stress. Researcher Phillip Marucha found that students who had oral surgery during exam time took 40 percent longer to heal than if they had the surgery during summer vacation. Their immune systems were compromised by the stress of exams.

It's not just our immunity that reacts to stress differently in the short term than in the long term. Our memory also considers adrenaline a helper and glucocorticoids a hindrance. Adrenaline causes additional glucose to be released in our bodies, and this gives our memory and mental acuity a boost. We think more sharply and remember things better. If you think of that primitive man with a bear attacking him, you can see that he had to be able to make razor-sharp observations of a threat and remember it well so that next time he could avoid danger. But, once stress becomes chronic stress, our memory suffers; glucocorticoids have been shown to actually erode memory over time. As Stephen Maren, a neuroscientist at the University of Michigan, says, "a brain system designed to keep us from being eaten is now eating away at us."

One of the most detrimental ways that stress eats away at us is by increasing our risk of cardiovascular disease, which comprises such ailments as strokes, heart attacks, and aneurysms. Cardiovascular disease kills more than twenty-six hundred people *daily* in this country, and has a mortality risk greater than the next six causes of death combined. And stress is one of the biggest contributors to this disease. Many experts, including Dr. Dean Ornish, a highly respected cardiologist, believe that controlling stress may be the single most important thing you can do to prevent heart disease. The fight or flight response causes our blood vessels to constrict and our hearts to beat faster and harder. This, coupled with the increased ability of the blood to clot during stress, sets us up for heart problems.

(After September 11, I often pointed out to clients that more people die every two days from cardiovascular disease than died on September 11. The real threat for the majority of us is not terrorism, but the stress it causes in our bodies.)

Chronic stress has many other negative effects. Here's just a sampling:

- Trouble sleeping
- Stomach and digestion problems
- Possible increased cancer risk
- Higher risk of autoimmune diseases (e.g., lupus, MS, thyroid conditions, rheumatoid arthritis)
- Higher rate of depression or anxiety

In fact, it is estimated that 60 to 90 percent of all visits to primary care doctors are for stress-related problems. Chronic stress can make us physically sick, and it can also bring on emotional distress as well.

POST-TRAUMATIC STRESS DISORDER (PTSD)

The crises of the last few years have made us all familiar with PTSD. The disorder actually became part of our lexicon in 1980, when it was first included in the American Psychiatric Association's bible, *Diagnostic and Statistical Manual* (DSM). PTSD was classified as an anxiety disorder because of the prevalence of signs and symptoms in Vietnam veterans. But war stress was not a new thing. In World War I it was called *shell shock* and in World War II, *battle fatigue.* But the actual classification of it in the DSM made it easier to diagnose.

After September 11, most of us assumed that emergency workers and those who were in the buildings that were attacked might suffer from this disorder. What I began to see in my broadcasting clients after September 11, however, was that many of them were also describing symptoms that indicated more than simple stress from the event:

Symptoms of Crisis Stress

Sleeplessness

Irritability/anger

Headaches

Crying easily

Withdrawing from people

Fearfulness

Feeling hopeless

Exhaustion/fatigue

Eating more/less

Stomach problems

Increased use of alcohol

Increased use of tobacco

Difficulty concentrating

Difficulty making decisions

Overidentification with survivors' and victims' grief and trauma

Preoccupation with disaster: watching/listening to news excessively in off hours

Staying overly busy: for example, volunteering to work extra hours

And, for some clients, these symptoms were interfering with their ability to work at their normal level. They also reported more problems in their relationships and with their health. Some appeared to be suffering not simply from stress but from PTSD. This is how I gauged their level of stress:

 **Mild**	 **Moderate**	 **Severe**
Symptoms are noticeable but not affecting life or performance	Symptoms could be severe enough to interfere with work and/or relationships but for less than one month	Symptoms interfere with with work and/or relationships for more than one month

The first two classifications accounted for most of the clients I saw after September 11, but there were some clients who fit into the "severe" category. Their symptoms may have started immediately after September 11, or they may have come on later, as if they had been smoldering in the clients' minds for weeks. These clients were the ones whom I believed were suffering from PTSD, and I recommended to them that we work in conjunction with a mental health professional.

To be classified as PTSD, symptoms must exist for a month or more. One of the main markers of PTSD is difficulty sleeping or nightmares. If these occur with regularity and keep you so exhausted you notice it during the day, you could be dealing with the disorder. Another common symptom is emotional upheaval. You may be dealing with PTSD if, when something reminds you of the trauma, you find yourself more irritable or suffer bursts of anger. You may have days where sadness and tears haunt you. Coupled with these symptoms, you may also have difficulty concentrating. Images of the crisis in the form of flashbacks may fill your mind when you should be working or relaxing. You may also find that you are jumpy. Even the smallest noise may startle you. You may be more aware of things around you as you become hypervigilant, like the primitive man in the forest always looking for the next threat.

After September 11 many people found that there were triggers that brought on the symptoms of PTSD. Any sensate reminder, like the smell of smoke or the sound of a siren, might trigger memories of the event.

Fatigue and stress can make a person more susceptible to these triggers. These memories can come slowly into your mind or they can explode before you in the form of a flashback. Flashbacks can be the most difficult to handle because they are so vivid and come on so suddenly. When you're in the grips of a flashback it's as if you are experiencing the trauma. It doesn't feel like the memory it is.

Many people found it hard to watch the television images of September 11 after that day because it brought up so much stress for them. And anniversaries of traumatic events almost always bring back memories of the event along with the stress of it. All the reporters who wrote about September 11 for this book mentioned the stress of the six-month and one-year anniversaries.

Risks for PTSD

These reporters' experiences after September 11 fit with the usual aftermath of a crisis. Glenn R. Schiraldi is an expert on PTSD and has written a sourcebook on the topic (see Suggested Readings). He explains it like this: "Most people will experience at least some symptoms of PTSD following a severely stressful event." In truth, the symptoms of PTSD are normal reactions to an abnormal event. A catastrophic stressor is outside the range of usual human experience. Our minds are not equipped to handle these catastrophes easily.

But everyone's brain chemistry is different. For example, studies show that only about 20 percent of soldiers exposed to combat will develop PTSD. This is the group for whom the memory of the traumatic event never dims. Their brain chemistry actually causes the memory to be recorded differently. But you don't have to be fighting in the battle to develop PTSD. A study done by Anthony Feinstein that appeared in the September 2002 American Journal of Psychiatry showed that war *reporters* suffer rates of PTSD at the same levels as combat veterans. (See *www.dartcenter.org* and *www.rsf.org* for comprehensive information about PTSD and reporting.)

There are certain risk factors that make a person more likely to develop PTSD. Some of these are things we can control and some of them aren't:

- Prior trauma such as physical abuse, sexual abuse, rape
- Lack of resilience or low self-esteem
- Difficulty expressing feelings to others
- Pessimism or depression
- Genetic factors
- Drug abuse

In my experience, one of the most significant factors is a prior history of trauma. There may actually be PTSD symptoms that are lingering from a prior trauma that become activated by a new crisis. If the initial trauma has not been processed, it can increase the severity of PTSD later on.

What Causes PTSD?

Events like September 11 seem like obvious causes of PTSD, but there are many others. And many of the other causes are events you may cover on a regular basis. Look over the list below, which I have summarized from the DSM criteria for diagnosis, and check off the events you've covered lately. You don't have to have been the victim. The PTSD diagnostic criteria say that a person does not have to have experienced the trauma himself. Remember that war reporters have been shown to have the same rate of PTSD as soldiers. What triggers PTSD are the intense feelings of fear, horror, and helplessness that the event provokes.

Events That May Cause PTSD

Check off any events you've experienced or covered for your job:

Catastrophic Events

- ❑ Combat
- ❑ Physical or sexual abuse
- ❑ Torture
- ❑ Assault
- ❑ Violent crime
- ❑ Viewing severed body parts
- ❑ Terrorism
- ❑ Riots
- ❑ Sniper attacks
- ❑ Cult abuse

- ❑ Bombings
- ❑ Death threats

Accidental Disasters

- ❑ Traffic accidents
- ❑ Fires
- ❑ Explosions
- ❑ Buildings collapsing

Natural Disasters

- ❑ Hurricanes
- ❑ Tornadoes
- ❑ Floods
- ❑ Earthquakes
- ❑ Fires
- ❑ Animal attacks
- ❑ Sudden deaths

My guess is that if you have been a broadcaster for any length of time you have checked off over half of these events as things you've covered. Reporting on these events is a normal occurrence for most reporters and anchors. As you'll see in Desirée Berenguer's profile (see chapter 14), you can be exposed to crisis stress when covering what most people would consider ordinary stories.

There are certain conditions that make a stressful occurrence more likely to cause PTSD. According to Glenn R. Schiraldi, they include:

- Unpredictability
- Long duration
- Recurrence with limited time to recover
- Involvement with multiple traumas (e.g., a building collapses with people dying, body parts visible, and fires)

If you think of an event like the D.C. sniper, you can see how it was a probable trigger for PTSD. No one knew when the next shooting would happen. The crisis lasted almost a month. There was little time to recover from reporting on one shooting before another one happened. In fact, six people were killed in the first twenty-seven hours. All of these conditions made this event very likely to cause PTSD in susceptible individuals.

It's the intentional human acts, like the sniper or terrorism, that are likely to produce the most severe cases of PTSD. Fortunately, not all traumatic events are as difficult to recover from. Natural disasters have been shown to be the quickest from which we recover—perhaps that's because they are truly beyond our control. But even a severe case of PTSD usually improves considerably within three months when the sufferer works through it with a mental health professional. I refer clients to psychologists and social workers who specialize in PTSD. Treatment options and other helpful information are available from the Department of Veterans Affairs' National Center for Post-Traumatic Stress Disorder Web site *(www.ncptsd.org).*

Without help dealing with PTSD, the memories, flashbacks, and stress can continue for a lifetime. There is no shame in seeking help from a mental health professional if you suffer from the symptoms of PTSD. It is not a mental illness. You are not crazy or weak or even unique. Your symptoms are shared by thousands of people who have lived through catastrophic events. PTSD is one of the things that can happen when a person is in a traumatic situation, whether they are a victim or an observer. PTSD is a condition that is caused by biochemical changes in the brain because of its reaction to overwhelming stress. This is not something that we can control any more than we can keep our leg from breaking if enough pressure is exerted on it. Seeking treatment for PTSD is not seen as a sign of weakness or a reason for your job to be in jeopardy. With help, PTSD does not have to affect your life forever.

CONTROLLING STRESS

As we have seen, however, stress does not have to reach the level of PTSD to have a dramatic impact on our lives. In fact, there are times in your career as a broadcaster when you need stress to help you perform better at your job. When you're covering a breaking news event as a reporter, stress is your ally. During a riot or protest demonstration, for example, you will need the fight or flight response to help you do your job well and avoid getting hurt in the process. If you're a news director or producer, or you're on the desk, you may need some adrenaline to help you adjust your broadcast to a late-breaking story.

But these events should be time-limited. Once the event is over, you need to be able to rest and relax. Rest and recovery from stress are what make it less harmful. When there is time for the glucocorticoids to leave your body, you can return to normal. A normal stress event allows for this (see figure 2-1). But, when you have no time to recover from the stress response, it changes from something that is helpful and, at times, lifesaving, into a destructive and possibly life-threatening disease. Unfortunately, many days in the newsroom can look like a stairway of stress with no recovery time.

Unpredictability is one of the cornerstones of news broadcasting. For many journalists, it's what they love about this business. They never know what will happen when they show up for work. They do know it will be a day filled with split-second decisions, constant deadlines, and possible crises. But this very unpredictability is one of the aspects of stress that makes it the most detrimental. In one scientific study, for example, rats that were given a warning bell prior to being shocked developed fewer ulcers. It was the rats that never knew when the shock was coming that suffered the most. When there is a lack of control, our stress response has a hair trigger. It may go off when the stress is not actually there. We have to be ready to spring into action instantly, and always prepared for the worst.

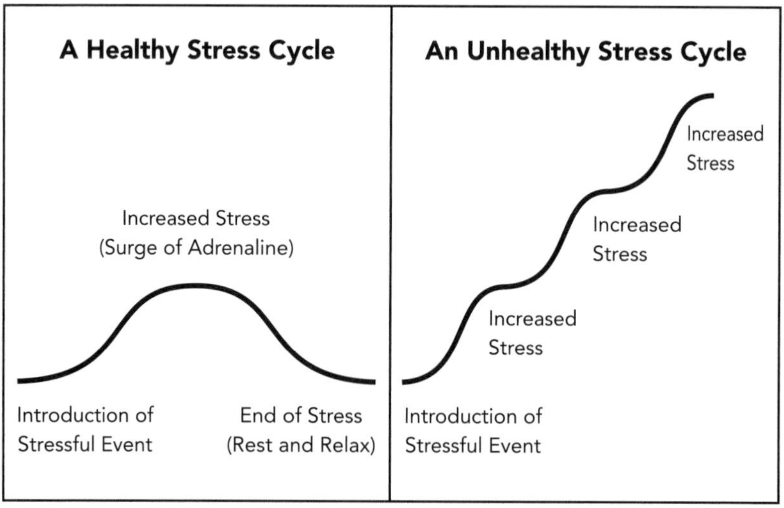

Figure 2-1

Think of the unpredictability Melinda Murphy faced as she looked down on the events in New York City on September 11 (see chapter 10). She didn't know what would happen next. Mike Walter certainly dealt with unpredictability all day on September 11 as more warnings came in that planes were headed toward Washington (see chapter 9). Now think of what you face daily in your job. This is not a profession that offers much predictability at all. And our bodies are not designed to deal with unrelenting, unpredictable stress. The fight or flight response is designed to help us fight off that bear. It's not designed to cope with the stressors of the news business.

To escape the emotional and physical stress of a crisis, the first impulse most of us have is to numb ourselves in some way. This may be by using alcohol or drugs or over- or undereating. Overwork is another very common way that I observe broadcasters coping with crisis stress. Work blunts our emotions and gives us little time to think about what may be bothering us.

The problem with using any form of numbing is that it only works in the short term, and at a high cost. When you numb the bad feelings you have, you lose the good ones as well. If you overwork to escape the memories of a stressful experience, you leave no time to enjoy life. Numbing negates any possibility of self-care.

Numbing also affects your ability to be compassionate and connected to the stories you report. Remember that just because you're covering the hundredth fatal traffic accident of your career doesn't mean that it is not a major catastrophe for some family. Think about how many times a veteran airline pilot has flown into a major airport. Does that give him permission to land your flight without his full attention?

Instead of becoming numb, we might instead become obsessed with the event, as Mike Walter did after September 11 (see chapter 9). This leaves no time for us to have the balance in our lives that offers relief. We turn away from our friends and family so that we can continue to focus on the trauma. We may feel that no one else will understand unless they've been through what we have. This can lead to burnout pretty quickly.

Negative coping activities like obsession or numbing only postpone any real improvement, and they often create new problems for you to face. Blocking out the horror of an event can't last forever. Eventually the feelings have to be dealt with.

What works is to accept what has happened and seek relief. In the next chapters you'll see how you can use RELIEF to deal with crisis stress. For most stressors that do not progress to PTSD, this is all you need to do. And, even if you have developed PTSD, the steps of RELIEF will help you in your recovery.

CHAPTER

3

R = Relax

"Everyone thinks of changing the world. No one thinks of changing himself."

Leo Tolstoy

The idea of considering relaxation during a crisis situation seems like a contradiction. The situation often feels so overwhelming that relaxation has no place in it. But a crisis may be the exact time to focus on relaxation. As Vietnamese Buddhist monk Thich Nhat Hanh says, you may need to think of "taking a break, to avoid a breakdown."

We all know you can't start a car that has a dead battery. There's no energy to draw from that battery. But during a crisis many of us ignore this same fact about ourselves; we try to keep going when all our energy is depleted. If one of these statements sounds like something you'd say, then relaxation is what you need:

- I feel burned out.
- I haven't had a minute for myself.
- I'm meeting everyone else's needs and not my own.
- I've been rushing around for days.
- I'm running on empty.
- I've lost touch with who I am.
- My life is out of control.

One common theme in all these statements is that life feels out of control. A relaxation break is an activity that not only calms you down, but also gives you a sense of having control again.

Relaxation breaks can truly be the preventative medicine that keeps you going during a crisis. Shortly after the attacks of September 11, Rudy Giuliani, the mayor of New York at the time, was asked how he got through those first awful days. He said, "My father used to say to me, if you are in a crisis and everyone is getting very emotional, you have got to become calmer. That is the only way to get yourself through."

Relaxation is also standard therapy for anxiety. Learning to incorporate relaxation into your daily life has been shown to help prevent a stress buildup that can result in a panic attack. Doing some relaxation every day is like putting out brush fires before they build into a firestorm. The firestorms of stress are panic attacks and other life-threatening health problems.

But how do you achieve that calm in the midst of chaos? During a crisis you can't run to a spa or take a vacation or walk away from what you have to cover. What you can do is find ways to relax no matter what is going on around you. Relaxation is simply taking a few minutes (or even a few seconds) to move from where you are mentally to another place. I call it going from Point A to Point B so that you can return to Point A refreshed. This doesn't require leaving where you are physically. This is a mental journey. And the more often you take this mental journey, the more you will be able to respond to crisis situations with a new sense of calm. You can return to work as the best of who you are.

This doesn't mean that you deny what is going on or take a Pollyanna view of it. It means that you take a mental and emotional break. Andrew Lack, who was president of NBC during the attacks of September 11, summed this up quite well when asked how his staff had been told to get through the stress of the event: "Our answer at NBC News is simple. Get used to it. Get over it. Take a breath. Take a step back."

The sobering reality is that, if you don't make time to relax, your body will often force you to. For many people, sickness is the only thing that stops them long enough to relax. They're literally forced to go from Point A to Point B because they're flat on their backs. It's also true that, if we don't find healthy ways to relax, we often turn to unhealthy ones like alcohol, food, or zoning out on junk TV. None of these provides true relaxation.

Fortunately, there are three simple ways to relax that can be used at any time and in any place. I advise clients to make time for these techniques at least once an hour. Each of these relaxation breaks can be as short as a few deep breaths, which might take less than thirty seconds. These breaths can be used to bring a sense of calm at the anchor desk, or on the battlefield during war coverage. All the relaxation sessions are built on the foundation of focusing on breathing. Once you learn to use your breath to relax, you can add calming, affirmative statements and visualizations to increase your sense of relaxation.

TECHNIQUE 1: BREATHING TO RELAX

Obviously you know how to breathe or you wouldn't be alive. But you may not know the impact your breathing has on your stress level. You have already read how being stressed speeds up your heart rate and increases your blood pressure. Deep breathing does the opposite—it reverses the stress response. Taking a deep breath can slow your heart rate by ten beats per minute or more. This can be a lifesaver in a crisis.

To experience how your breathing is affected by stress, make a tight fist with one of your hands. Notice that you stopped breathing when you did this. Excessive tension in the body usually causes breathing to stop for a short time, and extended stress may speed up breathing and make it shallow. Both of these responses are part of the fight or flight response, and they make it difficult to relax.

Rapid breathing is also very contagious. We often match our breathing pattern to those around us. So if you're in the midst of a crisis you will most likely increase your stress by copying the rapid, shallow breathing of others around you.

To combat all this rapid breathing, you have to consciously slow down your respiration. First let's look at how to produce a relaxing, deep, abdominal-diaphragmatic breath (one that involves the abdominal and diaphragm muscles). Proper breathing should begin low in the torso around your waist. You should feel your stomach expand as you inhale and contract as you exhale (see figure 3-1). One way to feel this breathing is to imagine that you have a beach ball right under your rib cage. As you inhale, fill up the beach ball. As you exhale, let it deflate.

Various postures and movements can help you refine this technique, because they force abdominal-diaphragmatic breathing. Try these activities and focus on the movement around your abdominal area, your sides, and your back. Some postures may work better for you than others. In all of them, concentrate on your breathing. Work with these exercises until you feel the proper movement:

1. Bend from the waist at a 90 degree angle, letting your arms and head hang relaxed. Keep your knees slightly flexed. Remain in this position until you can feel your abdominal-diaphragmatic breathing.
2. Sit forward in a chair and put your elbows on your knees. Breathe normally and focus attention on the location of the movement.
3. Tilt your head back and yawn deeply. Feel the movement in your abdominal area. This is also an excellent way to relax the throat.

One of the best postures for feeling abdominal-diaphragmatic breathing is lying down. Find a comfortable carpet or bed and stretch out on your back. Spread

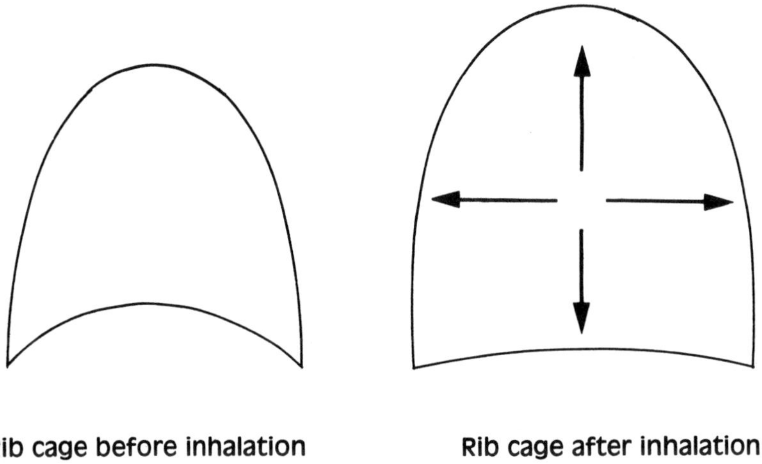

Figure 3-1

your legs slightly and move your arms away from your body so that there is open space in your armpits. Perform the following activities:

1. Close your eyes and concentrate on your breathing.
2. Place your right hand on your chest and your left hand on your abdomen. Notice that you can keep your right hand still while your left hand rises and falls with the breath.
3. Place a book on your abdomen and watch it rise as you inhale and fall as you exhale.

You can also work with a partner. Stand opposite your partner and put your hand on his or her abdominal area just above the waist. Ask your partner to inhale and attempt to push your hand away. Focus on the movement of the abdominal area. Switch tasks. If either of you has difficulty, forget about breathing and simply push the hand away with your abdominal muscles. Concentrate on moving those muscles, and then put an inhalation with the movement. (These breathing exercises and others are available in my book *Broadcast Voice Handbook*. See order information at the end of this book.)

One simple way to use the breath to relax is to take a few breaks during the day to focus on your breathing. Simply stop and *breathe!* Allow your breath to slow you down. Imagine that beach ball below your rib cage and feel it filling with air. Take five or ten deep breaths.

If you want to do more with your breathing to create a relaxed state, try this exercise. Take a slow inhalation through your nose, letting your abdomen expand as you inhale. Then gently exhale and imagine blowing out a candle. Purse your lips and slowly blow the air out; try to exhale a bit longer than you inhaled. Repeat this a few times. You may find it makes you a bit lightheaded at first. Always return to your normal breathing if this occurs. Using this blowing exercise will focus your thoughts on your breathing. This can work as an instant relaxer.

If you are in the middle of a press conference or preparing for a live shot in front of a crowd of people, you might feel a bit silly blowing out an imaginary candle. What you can do in this setting is simply to yawn. A good yawn increases the oxygen in the body, relaxes the jaw and mouth, and has a calming effect. And it's a fairly socially acceptable thing to do in most situations. Remember the value of a good yawn.

TECHNIQUE 2: RELAXING, AFFIRMATIVE THOUGHTS

During a crisis, information is usually pouring in at a rate that is difficult to handle. In fact, information overload is a big stressor during a crisis. And in some ways it's easier to deal with lots of chaotic thoughts than it is to free our mind of thoughts. In Mike Walter's profile he talks about how he felt like he had an IV in his arm the first few days after September 11, and it was supplying him with a steady stream of information. He wanted all the information he could get. One of my clients told me during that time her newsroom felt like "all trauma, all the time."

If you persist in holding onto stressful, negative thoughts, they will increase your stress level both physically and mentally. When Mike Walter was sitting home in his comfortable, safe living room thinking about the Pentagon tragedy or watching the news coverage, his body did not know that he was not still at the scene. The stress response does not discriminate between *thinking of* a stressful event and *being in* a stressful event. For all Mike's body knew, the plane was crashing over and over again. He was experiencing it physically every time he replayed it in his mind.

A perceived threat is just as real for us as the actual threat; our thoughts cause biochemical changes in our bodies that communicate with every cell. Dr. Robert S. Eliot, a stress expert, points out that stress literally "begins in the *mind* but ends in the *body.*" When we think about stressful events, our body pays the price.

(If you question whether your memory is this closely connected with your body's reactions, consider the last time you had goose bumps. This reaction is part of the fight or flight response. In animals it causes their fur to bristle, which often makes them appear to be larger and, thus, more formidable opponents. For us, it's a good way to feel the mind affecting the body. Think of a time when you were really cold, actually shivering, or remember a point at which you were suddenly frightened. If you can do this completely enough, your mind will signal your body to respond, and you will get goose bumps. This is another way the mind can signal the body to react.)

So how do you combat these stressful thoughts and memories? One way is to couple your deep breathing with relaxing, affirmative statements. Stress researcher Dr. Hans Selye tells us, "Nothing erases unpleasant thoughts more effectively than pleasant ones." Since we can only hold one thought at a time, using your relaxation period to replace the stressful thoughts with positive ones can be very helpful. And an extra benefit is that you can train your mind to go to the positive instead of the negative by using repetition. Repeating positive phrases can have the same effect as that popular song you hear so many times that you can't get it out of your head.

The relaxing, affirmative words you repeat can be anything you like. It can be as simple as repeating the word "relax" as you inhale and exhale. Coupled with deep breathing, this can move you into a relaxed state.

Try repeating "re-" as you inhale and "-lax" as you exhale. Take a deep inhalation and think "re-" and blow the air out as you think "-lax." Or try one of the phrases below:

- I'm okay now.
- I'm safe now.
- Life is not an emergency.
- Whatever comes my way, I can handle it.
- Remember to breathe.
- Slow down.
- Nothing to fear.

This is just a sampling of phrases you can use. Many people like to use a spiritual or religious word or phrase, which can add another positive component to your relaxation time. Whatever you repeat, let it take you to a positive place, and use your deep breathing to calm your stress response. If you find you have trouble at first keeping your attention on your word or phrase, don't get discouraged. It's natural for the mind to wander. Just gently bring it back to your word or phrase. These intrusive thoughts are simply cues to get you back to relaxing.

TECHNIQUE 3: VISUALIZATION

When we're having a lazy, unstressful day we often daydream. That's exactly what visualization is. It's conscious daydreaming or escaping into a scene in your head. PET scans have actually shown that when we think of pleasurable things it changes our brain activity. There's no doubt that imagined scenarios have the power to change our physiology. If you want proof, just think of the last time you had a nightmare. During this nightmare, your breathing most likely sped up. You began to sweat, and you may have fought off an imaginary attacker by moving your arms and legs. Your thoughts produced a very real fight or flight response.

In the same way that a nightmare can create stress, a peaceful vision can reduce it. You can use your relaxation time to escape into a memory or imaginary scene that will slow down your heart rate and calm your body.

Remember your last trip to the beach or the countryside? How many senses can you involve in this memory? Let yourself smell the air, feel the breeze, hear the sounds, and see the scene. Take an imaginary trip back there. Remember to use deep breathing while you're on this trip.

Just this simple daydream can be a great stress reliever. One thing that Sherry Stalley used after September 11 was a photo of her children, which she tucked into her purse that first night (see chapter 11). When the stress of covering the aftermath of the crash of Flight 93 became too much, looking at this photo took her away from the events around her for a short time. Photos, letters, pictures, and other things of meaning can help with your visualization. All I have to do is look at a photo of my cat when he was a kitten to escape into a pleasant, unstressed place.

One way to begin to find some possible images to use is to write down all the things you can think of that bring you pleasure. This might be your dog or cat's warm fur, a juicy peach, a pizza, a warm blanket, your loved one, or anything that comes to mind. You might also include all the things you can think of that you're grateful for. Let this list be your guide when you begin a visualization. Simply breathe deeply and spend a few minutes focusing your thoughts on one of the things on your list.

LONGER RELAXATION

Short relaxation periods can be a part of even the most stressful day. During a crisis, they may be the only relaxation you can get. But, when you have the time, look for ways to relax for longer periods. This is a great antidote to the chronic stress of broadcasting. You may already be thinking of things you like to do that are relaxing and enjoyable. I know that if I can spend even ten minutes in the hammock on my deck I will be relaxed. Looking at the trees, the sky, and the birds will definitely take me to Point B no matter how stressed my day has been.

One client of mine lives in New York City in a high-rise condominium. She turned her small balcony into a beautiful garden with potted shrubs and flowers. A few minutes tending her garden is all she needs to relax at the end of a hectic day. Another client working in a more rural area bought a horse. She told me that just the thought of going to the stable to be with her horse would relax her in the middle of her workday. Here are some other ideas for relaxing that clients have shared with me:

- Running
- Skiing
- Swimming
- Golfing
- Walking
- Getting a massage
- Doing anything creative like painting, needlework, or making pottery
- Coloring in a coloring book
- Playing with children
- Petting an animal
- Photography for pleasure
- Reading a novel
- Listening to relaxing music
- Taking a long, hot bath or shower
- Baking or cooking for pleasure

The one thing all these activities have in common is that they take your mind off what's going on in your life. They allow you to become fully absorbed in doing the activity. Spacing out in front of the television or watching a movie may not do that. We often become more stressed during these events—depending, of course, on the content. Remember that you want to reach a calm state when relaxing. You want your mind to stop racing. Most television and movies excite rather than calm.

But you can't force yourself to relax. To illustrate this point, one year at the Radio-Television News Directors convention I actually entitled a workshop I gave "Relax, Dammit!" Forcing relaxation only makes us tenser. As author Martha Beck says, "This is like trying to put out a fire with gunpowder" (see Suggested Readings). For relaxation to be effective, we have to give ourselves permission to relax. If we do this not only will we create positive physiological and chemical changes in our bodies, but our minds will also receive a break from the chaos that may surround us. No time could be better spent.

The Bottom Line of Relaxation

1. Relaxation takes you from where you are (stressed) to another place.
2. Deep abdominal-diaphragmatic breathing is the basis for relaxation techniques.
3. Relaxing, affirmative phrases can be added to the breath for a deeper relaxation.
4. Visualization, or conscious daydreaming, offers an escape from stress.
5. Longer relaxation periods can enhance your daily life.

CHAPTER

4

E = Exercise

"Aerobic exercise is a stress that makes us tough enough to handle all other stresses."

Covert Bailey

Now that you have a sense that you can fit some relaxation into your day, you probably have questions about the possibility of exercising during a crisis. When news is breaking every second, you obviously can't leave what you're covering to go the gym or jog three miles. But fitting exercise into your day when you're in a crisis situation is imperative to combat the rising stress level.

Exercise is not just a necessary part of dealing with a crisis. It is one of the most important things we can do for our health. Imagine how eager you would be to take a pill that provided all these benefits:

- Gives you energy
- Improves your immune system
- Reduces feelings of anxiety and depression
- Improves sleep
- Builds healthy bones
- Reduces risk of heart disease and cancer
- Lowers cholesterol and blood pressure
- Reduces risk of stroke and diabetes
- Improves job performance
- Controls weight and body fat percentage

All of these benefits are provided by regular exercise.

Exercise is unquestionably one of the most important things you can do to keep yourself healthy on a day-to-day basis. Health experts now agree, virtually unanimously, that exercise is the single most beneficial thing you can do to promote good health. The *New England Journal of Medicine* reported in 2002 that poor fitness is a better predictor of death than smoking, high blood pressure, or heart disease. And the Cooper Institute for Aerobics Research found that regular exercise improves your immune system and, thus, reduces your risk of death by disease by 60 percent. This benefit alone should convince you to begin exercising.

Most of us ignore our cardiovascular health unless we've had a problem with it. Heart attacks seem like a remote risk. We worry more about colds and flu, because they can interfere with our day-today ability to stay on the job in a crisis. But the stress of covering a crisis can have a negative impact on the cardiovascular system. Exercise enhances your cardiovascular health enormously. That's important because more people—men and women—die from heart and blood-vessel diseases each year than from all other causes of death *combined.*

Another advantage of exercising during a crisis situation is that it gives you that much-needed sense of having control over something. You're in control when you exercise, no matter what chaos is going on around you. Just as taking time to relax gives you this control, exercise does as well.

But you may still be wondering how you can possibly work exercise into your day. Incorporating exercise into your life when a crisis is happening requires a new way of defining exercise. There are two kinds of exercise that help you deal with crises. One helps you get through a crisis and the other helps you prepare for crises before they happen.

STRESS RELEASE EXERCISE

Stress release exercise helps you on the spot. It works to lower the level of stress hormones in your body almost instantly. You will remember that the fight or flight response is an automatic bodily reaction that prepares us for an attack. It is fueled by adrenaline and sustained by glucocordicoids. But what happens when you're stressed but you can't fight and you can't run? The glucocordicords build up in your body. Exerting yourself physically is an integral part of the fight or flight response; your body wants intense activity like it would get if it really were being attacked. To reduce stress, we need to give the body something that will fulfill this need.

Let's imagine that you're reporting on an event like Neal Augenstein covered during the D.C. sniper crisis (see chapter 12). You're spending days outside the police media location waiting for updates from the chief. You can't leave. You're aware that you're a sitting duck for the sniper if he chooses to single out a newsperson. Your stress level is high for eight or ten hours at a time. This is exactly when you need to use stress release exercises.

Stress release exercises can last anywhere from a few seconds to several minutes. This is one of those situations in which anything is better than nothing. But if the exercise lasts for a full ten minutes, we get a big payoff.

Performing an activity for ten minutes that makes us slightly winded releases beta-endorphins, serotonin, and other feel-good chemicals in our brains. These make us feel relaxed and drive down the stress hormones in our system. This short bit of exercise also energizes us, because it speeds up the blood flow to the brain, bringing much-needed oxygen to brain cells. A study at California State University found that just ten minutes of brisk walking yields up to two hours of increased energy.

What types of stress release exercises might work in a crisis? It depends, of course, on your situation. If you're stuck at a site as Neal was, your choices are limited. But, if you can escape for ten minutes, here are some things that might work:

- A brisk walk or jog
- Climbing stairs
- Jumping rope
- Karate or kickboxing moves
- Any combination of the above

But what if ten minutes is out of the question? Remember that any activity is better than none. Performing any of the exercises above for even a few minutes will help. You can run in place for a minute or two. You might find a private place to throw a few punches into the air to fight off the stress. Anything that will simulate the fight or flight activity will give you the needed release.

Isometric Exercise

You can also use isometric exercises for stress release. These exercises pit one muscle against another or against an immovable object. I often recommend isometric exercises to anchors sitting at the anchor desk. They are fairly unnoticeable, and yet they can provide a good contraction of muscles. This contraction relaxes the muscle once it is released. Typical isometric exercises might involve tensing your buttocks while you're sitting or standing. Hold the contraction for ten seconds and release. You can also push against a wall or desk with your hands to contract your shoulder and arm muscles. Again, hold for ten seconds and release. Even something as simple as making a fist for ten seconds can help release tension when you relax your hand. You can also push your palms together to tense your arms and shoulders. Any of these moves will burn off some tension.

Stretching

Another important stress release exercise is stretching. This will not help you aerobically, but it will release tension. It will also improve circulation, reduce anxiety and fatigue, and improve your mental alertness. Stretching is important every day, whether you're covering a crisis or not. And it can be done almost anywhere at any time.

Stretching not only helps at the time you do it, but it also builds muscle memory of what a relaxed, stretched muscle feels like. This muscle memory will help you relax throughout the day. We have over six hundred skeletal muscles in our body, and each of these muscles can hold tension. Stopping to stretch these muscles all through the day will keep tension from building up in the body.

Let's look at a typical day for a reporter covering a crisis like Sherry Stalley did in the aftermath of September 11 (see chapter 11) and see how tension might affect her. She might clench her teeth while sleeping because of the residual tension from the crisis event, producing tightness in her jaw and face. She might hold tension in her forehead as she speeds back to the scene to do a follow-up piece. Her leg muscles might get tense while she stands waiting for an interview with an emergency worker. Back in the newsroom she may suffer shoulder and neck cramps while she writes at her computer. All of these muscles can hold onto tension for hours. Combine the tension held by each individual muscle, and you have a very tense body.

Stretching helps release tension so that it doesn't build up. An added benefit of stretching is that, if you do it right, you have to slow down and focus on your breathing. If Sherry put some stretching breaks into her day, she would also be adding relaxation.

It's never a good idea to stretch in a hurry, or jerk your body into a stretch, or bounce, or hold your breath while stretching. One way to avoid these mistakes is to concentrate on *feeling* the stretch. Clear your mind and let stretching take you to that relaxation point (Point B) where you forget about the stresses of your day for a few seconds. Hold each stretch for at least ten seconds. Be careful to stretch only within your comfortable limits. This may vary from day to day, but never stretch to the point of pain.

Here is a series of stretches that can be done separately or as a sequence. You can perform the entire series in less than five minutes. But you can also pick out a single stretch that might feel especially good, like shoulder rolls after working at a computer, and do that alone.

Do all these stretches in a standing position with your knees slightly flexed and your feet apart about as wide as your shoulders. Remember not to bounce or throw your body into a stretch. Hold each for ten to twenty seconds, feeling a gentle stretch. If you have any pain, do not do the stretching and check with your physician.

1. Begin with your hands at your sides. Slowly raise your hands above your head. Reach straight up several times as if you are trying to pick apples just above your reach.

2. Bring your arms straight out in front of you and interlace your fingers. Imagine someone is pulling your hands away from your body. Keep your chin parallel to the floor. Feel the stretch in your shoulders and upper back muscles.

3. Bring your hands behind your back with your arms straight down toward the floor. Interlace your fingers with your hands behind your buttocks. Try to pull your fingers apart, feeling the stretch across your chest and shoulders. Keep your chin parallel to the floor and avoid arching your back.

4. Pretend you are swimming the breaststroke. Begin with your arms straight out in front of you with the backs of your hands touching. Stroke back as far as you can comfortably with both arms. Repeat four times.

5. Pretend you are swimming the backstroke. With the same beginning position as #4, rotate one arm back and follow your hand with your eyes. Repeat with the other arm. Begin again. Repeat four times with each arm.

6. Roll your shoulders, moving both at the same time. Begin by pulling them up toward your ears. From this position, rotate them back so that your shoulder blades are coming together. Now stretch them down. Finish by rotating them forward as if trying to make your shoulders touch in front. Repeat this rotation four times. Change direction and rotate four times.
7. Place one hand on something for balance. Bend your opposite leg up toward your buttock. Grab the foot or ankle of the lifted leg and gently pull it toward the buttock. Feel the stretch in your upper leg and the front of your hip. Switch to the other leg when the stretch is complete.
8. In a seated position, bend forward over your knees so that the weight of your upper torso is resting on your thighs. Let your head and your arms hang in a relaxed, limp manner. Feel a deep stretch in your lower back. Push with your hands on your knees to come up slowly.

STRESS PREVENTION EXERCISE

Stress prevention exercise is what most of us think about when we plan to work out. It involves that thirty- to sixty-minute exercise session during which we train aerobically. We can do this by walking, jogging, jumping rope, exercising at a gym on any number of types of equipment, or attending aerobic exercise classes.

Basically, aerobic exercise uses the big muscles in our legs and buttocks to put a demand on our lungs and heart. The reason I call it stress prevention exercise is because it actually changes our body in a way that limits our stress level all the time. Covert Bailey, one of the nation's foremost health and fitness experts, explains it this way in his book *Smart Exercise:* If we're novice exercisers, our bodies secrete large amounts of stress hormones because the actual process of exercising makes the body feel like it is in the fight or flight mode. Think about it. If you've never been a jogger and you begin jogging, your body will think that you must be running away from a dangerous situation, that you're in grave danger. But if you continue jogging several times a week for a few months, your body realizes that you are not in danger. At this point it starts producing much less of the stress hormones. When it makes this adjustment for your exercise time, it makes it across the board. So, when you're in the middle of covering a crisis, your body will be secreting a lower amount of stress hormones if you are exercising regularly. This miraculous adjustment is one of the main reasons it's imperative to make aerobic exercise part of your life.

Health Benefits of Stress Prevention Exercise

If this physiological adjustment is not enough to get you exercising, look at the other benefits of aerobic exercise. Consider that it actually improves your immune system. Exercising increases your circulation, which pushes natural disease-fighting cells throughout your body. Think of it like a fire alarm going off in the firehouse. Instead of the firemen sitting around resting, they are up and running to the fire.

Exercise mobilizes the immune system in much the same way. It gets our disease-fighting cells circulating in our bloodstream and looking for disease invaders to kill.

Exercise also strengthens our heart muscles. A Harvard University Medical School and School of Public Health study found that women who exercised as little as *one hour per week* had about one-half the risk of developing heart disease as non-exercisers. In fact, this exercise provided as much benefit as giving up cigarettes or losing weight. That's an enormous payoff for very little exercise.

Exercise and Depression

One of the most frequent problems with covering crises is that broadcasters become anxious and depressed either during the event or after it ends. Exercise also combats these situations. Consider these studies:

- A 2000 Duke University study found that exercise is as effective as antidepressant drugs for mild depression.
- A University of Manitoba study showed that as little as twenty minutes of walking or jogging a day produced significant gains in the treatment of mild to moderate depression.
- In 1999 the University of Colorado published four animal studies showing that moderate exercise can reduce stress.
- A University of Wisconsin–Madison study found that the calming effect of exercise lasted up to five hours.

There's no disputing the fact that exercise combats stress and depression. The natural feel-good chemicals that it releases into our brains give us the mental boost we need.

How Much Exercise?

So what is the best recommendation to reap these benefits? The American College of Sports Medicine recommends twenty to sixty minutes of aerobic exercise at least three times per week (make it five to seven times per week if you're trying to lose weight). The easiest gauge of what is aerobic is that you should be winded when performing it, but not so winded that you can't carry on a conversation. This exercise, however, can be performed in ten-minute sessions throughout the day. It doesn't have to be one continuous training session. (Note: If you have not been exercising, it is a good idea to check with your doctor before beginning any exercise plan.)

It's also recommended that you add weight and flexibility training two to three days per week. (Do not work the same muscles in weight training two days in a row. Give your muscles a chance to rest and recover for at least forty-eight hours.)

This type of stress prevention exercise needs to be part of your life. You can't wait until a crisis hits to start. Stress prevention exercise really needs to be a lifetime commitment, but the payoff for this commitment is well worth the time and effort.

You'll become a better stress-processing animal, and your health will benefit in almost every way.

When it comes to making exercise a part of your daily life, what works for people really varies. Many love the camaraderie of working out with other people. Others like to make exercise a personal time. If you love the gym, you'll usually find time to go there. But there are other options. One of the best investments you can make is to buy a piece of exercise equipment to have at home. With equipment at home, you can exercise anytime, day or night, and in any kind of weather. It doesn't matter if you buy a treadmill, stationary bicycle, elliptical machine, or any of the other kinds of equipment that are available. Add some free weights to your purchase, and you'll be able to meet all your exercise needs at home. This equipment will help you achieve your goal of working out on a regular basis.

Covert Bailey sums up the value of exercise very effectively when he uses the example of a car. Who wouldn't want to buy a car that automatically tuned itself up every time we drove it? Well, that's exactly what our bodies do when we exercise, because exercise makes demands on every organ in our bodies, and these organs improve with use. Exercise really works.

The Bottom Line of Exercise

1. Exercise is considered by most medical professionals to be the single most important thing you can do to maintain good health.
2. Stress reduction exercise involves doing something to release adrenaline and glucocordicoids from your body. It can be done in as little as ten minutes of walking, jogging, or other activity.
3. Isometric exercise works well for stress reduction.
4. Stretching should be included in your daily stress reduction activities.
5. Stress prevention exercise is aerobic exercise for thirty to sixty minutes.
6. Weight lifting and flexibility exercises round out the recommendation for a healthy workout.

CHAPTER

5

L = Let Yourself Sleep

"Not paying attention to sleep is tantamount to catastrophe."

David F. Dinges, Chief of Division of Sleep and Chronobiology, University of Pennsylvania

During World War II, commanders in the Royal Air Force noticed a strange occurrence. It seemed that many pilots fought well in battle but then mysteriously crashed returning home. Lack of sleep was found to be the culprit. In an article in the *Washington Post* entitled "In War, Fatigue Can Be a Deadly Adversary," Shankar Vedantam explains that the adrenaline of the mission kept the pilots alert during the actual bombing, but fatigue overtook them on their flight back.

This story resonates with me because I see so many clients who operate in the same way. When working a crisis, when their adrenaline is high, they are completely unaware of how sleepy they are. Once they start home, however, it hits them. Sitting behind the wheel of a car is not a good time to fight fatigue; the National Highway Traffic Safety Administration reports that driver fatigue results in more than one hundred thousand accidents per year.

Why do these accidents happen? A study published in *Occupational and Environmental Medicine* reported that people with too little sleep actually performed as if they had had two drinks (the equivalent of a blood alcohol level of 0.05). This affected concentration, motor skills, and coordination. You simply can't perform at your best when you haven't gotten enough sleep.

The first question I'm always asked about sleep is, *how much sleep do I really need?* Every study I've read says all of us need a minimum of seven hours per night. Some of us may need more. Disease, for example, can increase the need for sleep—as can stress.

WHY SLEEP?

The negative effects of lack of sleep aren't limited to traffic accidents. Researchers are finding that losing sleep may be as harmful to our health as not exercising or eating the wrong foods. Paul T. Gross, director of the sleep disorders center at the

Lahey Clinic, explains: "Studies have suggested that sleep deprivation can affect every aspect of how the body functions, from the ability to memorize and logically reason to the repair of neurons and development of muscle." At Brigham and Women's Hospital in Boston a study showed that a woman who gets six hours of sleep per night raises her risk of developing heart disease by 30 percent over a woman who sleeps eight hours. If sleep drops to five hours per night, that percentage jumps to 52 percent. The body simply can't maintain its healthy functioning when sleep is limited.

When we don't get the sleep we need, our immune systems suffer. Michael Irwin of the San Diego Veterans Affairs Medical Center reports that just one night of reduced sleep greatly reduces the activity of natural killer cells in our bloodstreams. These are the cells that fight off viruses like those that cause colds and flu. We also can't concentrate as effectively, and are generally more accident-prone. The reasons for all these effects are not known, but researchers do know that sleep plays an important role in our physical and emotional health.

One of the main ways that sleep helps us is that it allows many of our cells to shut down and repair themselves. Cells need this time to clean out cellular debris that accumulates when they are functioning. It also allows time for new cells to generate. Author Pamela Peeke (see Suggested Readings) explains this phenomenon with the following metaphor. Suppose the executives in an office building decided to work all night. The cleaning crew could not come in to clean and no maintenance could be done on equipment. If the next night they did the same thing, the trash would begin to pile up and nothing would be repaired. Eventually, there would be so much trash the executives could not work. Equipment would begin to break down because no repairs had been made. This is much like what happens to our bodies without sleep. The end result is possible illness and a rapid aging of our cells.

And we don't even have to stay up all night to suffer. Loss of a few hours of sleep can be a problem. Missing as little as an hour or two a night creates a "sleep debt." After one or two nights you may miss a total of four hours of sleep. This might result in fatigue and irritability. If you continue to lose an hour or two a night, you will eventually have a sleep debt that totals an entire night's worth of sleep. The sleep debt doesn't just accumulate; it compounds as well. This is when you raise your risk of illness and accident.

One of the insidious things about lack of sleep is that we often don't know when we're sleep deprived, especially if a sleep debt builds up gradually. The reason for this is that lack of sleep affects our cognitive functioning. Slow but steady sleep deprivation affects the brain's ability to realize that it is sleep deprived. We may not feel the crushing fatigue we suffer when we miss a whole night's sleep, but the steady loss is affecting our bodies in the same way.

REM Sleep

The stage of sleep that's most valuable to health and well-being is the REM (Rapid Eye Movement) stage. This is the point at which dreaming occurs. If you're having nightmares it may seem like they last all night, but we actually only dream in ten to twenty-five minute cycles a few times a night. Adults need for their overall sleep to include around 20 percent of REM sleep per night.

Experts generally agree that our sleep time has five stages. As we cycle through the stages of sleep, we go from stage 1, which is light sleep, to stage 4, which is the deepest level. Then we climb back up to lighter stages of sleep and finally to REM sleep. In REM sleep our eyes move, our heart rate goes up, and our blood pressure increases. The muscles in our limbs are temporarily paralyzed so that we won't act out the vivid dreams we're having.

How important is REM sleep? Well, studies have shown that rats, whose normal lifespan is two to three years, live only five weeks when deprived of REM sleep. Lack of REM sleep affects our ability to process what has happened during the day. The limbic system of the brain processes feelings, and this is the part of the brain that is most active during REM sleep. Certainly during a crisis we need this REM sleep time to help process the feelings that are arising.

REM sleep is also important for storing long- and short-term memories. When we're learning new skills, we spend more time in REM sleep. Infants actually spend 50 percent of their sleep time in REM sleep, possibly because they are taking in so much new information daily.

The only way to get more REM sleep is to get more sleep. If your sleep time is limited because you go to bed too late or get up too early, your REM sleep is limited as well. On the other hand, you may have good intentions to sleep enough, but you find your sleep time is interrupted. This will affect the amount of REM sleep you receive as well.

But all is not lost when it comes to sleep loss. Eve Van Cauter of the University of Chicago Medical School explains the solution: "Just as lack of sleep can harm the body, getting sleep can help it." Sleep comes with its own remedy. We can pay off our sleep debt with proper sleep.

STRESS AND SLEEP

There are three types of problems that interrupt sleep. You can suffer from *sleep onset insomnia,* which means you can't get to sleep easily; *sleep-maintenance insomnia,* when you wake up and cannot go back to sleep or wake up intermittently for a total of more than thirty minutes per night; and *poor quality sleep,* in which you never reach a deep level of sleep. Any of these types of sleep problems may occur in the aftermath of a crisis. After September 11 the National Sleep Foundation conducted a survey and found that half of all respondents reported some form of insomnia in the nights immediately following the tragedy.

The very nature of a crisis and the added stress that comes with it often make sleep difficult. In one of the workshops I gave shortly after September 11, I asked participants to tell me what their idea of "extreme self-care" during this stressful time would be. Over 50 percent of them said a few more hours of sleep each night. I pointed out to them that this request should not feel "extreme." Sleep is imperative when covering a crisis. You need to take care of yourself for the long haul, and getting at least seven hours of sleep a night is a must.

But, when a crisis hits, getting good, sound sleep is difficult. One of the reasons for this is that disturbed sleep is most often provoked by an inability to stop thinking about work. When that work involves a crisis, disturbed sleep often results. You may have trouble getting to sleep even though you feel exhausted, or you may be awakened by nightmares. A participant in a workshop I gave on stress control a few years back referred to these nightmares as "newsmares." He said he kept replaying the events of the day all through the night.

So how do you get the sleep you need during stressful times? Fortunately, you can make many small adjustments to your lifestyle, each of which will improve your sleep. The list below will help.

14 HEALTHY LIFESTYLE CHANGES FOR BETTER SLEEP

1. **Limit Your Caffeine Intake:** We all know that too much caffeine affects sleep. I advise clients to consume all their caffeine in the morning hours and limit caffeine intake to 200 mg or less. Caffeine can stay in your system ten to twelve hours, so try to avoid any caffeine for six to ten hours before bedtime. Also, be aware of all the sources of caffeine. Coffee is not the only culprit (see Crisis Survival Kit 5).

   Many of my clients are adamant that caffeine does not affect their sleep. What they may not realize is that caffeine will not necessarily make it difficult to get to sleep. You can build up a tolerance that will allow you to get to sleep easily after a few cups of coffee at dinner. But the problems arise because caffeine will keep you from reaching stage 4 sleep, which is the deepest level. It is at this level that cells are restored and your immunity is built up. The caffeine will take its toll without you being aware of it.

2. **Cut Back on the Smokes:** Cigarettes affect the REM sleep cycle, and smokers have reduced REM sleep. They are also more likely to awaken spontaneously after three or four hours, when the body goes into nicotine withdrawal.

3. **Watch Your Alcohol Intake:** Unfortunately, many broadcasters, like the general public, may turn to alcohol to help them sleep. It is estimated that 20 percent of all insomniacs rely on alcohol to find relief. The fact is, alcohol won't help in any meaningful way; in fact, it will do exactly the opposite. Initially, alcohol may relax you and help you get to sleep, but once your body has metabolized the alcohol you are likely to awaken and not be able to go back to sleep. It takes around three hours to metabolize alcohol, so more than a couple of drinks after dinner may affect your ability to go into REM sleep.

4. **Eat Early:** We all know that it's not easy to sleep on a full stomach. Digesting food can keep us awake. Going to bed within two hours of eating a big meal can also contribute to a condition called gastric reflux. This happens when the acid from the stomach comes up into the esophagus

and causes irritation. To sleep well and avoid gastric reflux, eat at least two hours before your bedtime. If you practice the grazing method of eating recommended in chapter 7, you'll avoid a large meal at night anyway and this will help you sleep.

Eating fatty foods close to bedtime is especially bad, since they take up to five to seven hours to digest. Fruit, on the other hand, takes less than an hour, and bread only one to three hours. If you must eat close to bedtime, make it a light, low-fat meal.

5. **Exercise at the Right Time:** Exercise is one of the best things you can do to keep yourself healthy, but exercise too close to the time for sleep can backfire by keeping you awake. In order to sleep well, body temperature should be on the cool side. Exercise raises your core body temperature, so exercise in the afternoon or up to three hours before bed, to give it time to cool down. A cooler body temperature means you'll get to sleep more easily and sleep more deeply. So exercise, by all means, but do it in the morning if you have no problems sleeping, or in the late afternoon or early evening if you want it to help you sleep.
6. **Keep Your Bedroom Cool:** A cool room can also lower your core body temperature and help you sleep.
7. **Set Sleep Times:** It is important to have a sleep schedule that you stick to as much as possible. We all have an occasional night when work or play keeps us up late, but this can wreak havoc on our ability to sleep well. Our bodies need to know when to expect to be awake and asleep. Irregular wake-up times actually disrupt sleep more than that occasional late night. Set a time to get up and stick to it even if you know you can sleep later. I have many clients who work morning shows, and the hardest thing for them is to keep their sleep schedule on the weekend when they don't work. No one likes to go to bed at 8 P.M. and get up at 3 A.M., but sticking to this schedule even on non-workdays is much easier on the body.
8. **Design a Sleep Area:** Most sleep experts agree that to make your bed conducive to good sleep you need to use it only for sleep and sex. This means no laptop computer, no phone calls, no reading except possibly something light, and no watching the news. When you sleep, keep the room as dark as you can. If you need a light, make it a small nightlight, since exposing your eyes to light makes your body think it's time to get up. Also, keep it as quiet as possible. If noise is a problem, consider wearing earplugs.
9. **Learn to Ramp Down:** You may remember when, as a child, you were told that it was time to prepare for bed. This might have meant a nice warm bath or a glass of warm milk. Maybe you were read a story. This kind of routine works for adults, too. Set a schedule for yourself that allows you to ramp down your activities. It doesn't work to engage in mentally active endeavors before bed. A client of mine who works the night shift told me recently that she couldn't get to sleep until two or three in the morning. She said she loved to get on the Internet and chat with friends or do

research when she got home around 11 P.M. This was clearly not a calming way to unwind.

10. **Get Out of Bed:** If you awaken and can't go back to sleep in thirty minutes, get out of bed. Do something relaxing like reading a novel or watching a non-action movie. Don't look at the clock. In fact, if you find you are waking up a lot in the night, cover the face of your clock to keep from creating stress. You might also try some of the foods below.
11. **Eat Sleep-Inducing Foods:** Certain foods contain tryptophan, a sleep-promoting amino acid. These foods include milk, cheese, turkey, chicken, and tuna. If you add some whole wheat bread to these, it will increase the absorption of the amino acid. Carbohydrates in general aid in the release of serotonin in the body. Serotonin regulates our moods and can help with sleep. Some good carbs to try include whole wheat products, sweet potatoes, oatmeal, and brown rice. Stay away from sugar, since it stimulates the adrenal glands and that wakes us up.
12. **Set Aside a Worry Time:** If you find that as soon as your head hits the pillow the worries begin, find a time to worry before bed. You might want to write down the things that are bothering you. Make a list or journal about them (see chapter 6). Do this several hours before bed and then put them away and forget them. If you wake up consumed with worries and can't get back to sleep, write them down and put them in a drawer. Tell yourself that you'll deal with them in the morning after a good night's sleep.
13. **Avoid Sleeping Pills:** The only time it's safe to take sleeping pills is if your doctor has prescribed them for a few nights. Taking sleeping pills for an extended time can result in rebound insomnia, which can affect your sleep for a long time. Sleeping pills also limit REM sleep. Use them only in an emergency.
14. **Don't Take Antihistamines:** If you use over-the-counter sleep aids, you're using antihistamines to get to sleep. These can dehydrate your body and, most importantly, your throat, which can be harmful to your vocal folds. Ear, nose, and throat specialists have cautioned me to tell clients to avoid antihistamines unless they are prescribed by a physician for allergies.

NIGHTMARES

One of the most difficult aspects of witnessing a catastrophe is that the sights often literally come back to haunt you. It's not unusual to have nightmares immediately after an upsetting event. In *The Post-Traumatic Stress Disorder Sourcebook,* Glenn R. Schiraldi explains that nightmares "are just the brain's attempts to process walled-off memory material" (see Suggested Readings). When something is too terrifying to deal with, we sometimes store it away. We want to avoid thinking about it during our waking hours, but we can't control what happens while we're sleeping. As Mike

Walter says in his profile, the images "hide in the recesses of my mind, popping up from time to time like an uninvited guest who refuses to leave" (see chapter 9).

The type of nightmares that often accompany PTSD are usually the worst. Schiraldi explains, "They can jolt people awake in a cold sweat, and feel so petrifying that sleep is avoided for fear of having another." In most cases nightmares this terrifying happen right after the event. As in Mike Walter's case, they usually gradually improve. A typical progression is for the first few nights of nightmares to be the most realistic and upsetting. With time, the nightmares become more metaphorical and less terrifying. As with any PTSD symptoms, if they persist for more than a month, seek help from a physician or mental health professional.

If you experience any sleep disturbance that does not respond to the suggestions in this chapter, and it becomes so great that it's making your job especially difficult, you may want to consider contacting an accredited sleep disorder treatment center. You'll find a list of these at *www.sleepfoundation.org.*

The Bottom Line of Sleep

1. A minimum of seven hours of sleep is recommended.
2. Fatigue affects our concentration, damages our immune systems, and increases risk of accidents.
3. We need sleep to repair our neurons and clean away cellular debris.
4. Long-term loss of an hour or two of sleep per night affects the brain's ability to realize it is sleep deprived.
5. REM sleep, during which we dream, helps build short- and long-term memory and assists in learning.
6. Consider the 14 Healthy Lifestyle Changes for Better Sleep to improve your sleep.
7. If nightmares affect performance during the day, seek help from a physician or mental health professional.

CHAPTER 6

I = Identify Your Feelings

"The most damaging feelings are those that are never discussed."

Dr. Don R. Catherall

If you have been through a few stressful or even traumatic events in your life, you probably know that trying to deal with irrational events using your rational faculties does not work. At some point, you have to pay attention to your feelings. Showing feelings is not a sign of weakness. It's a normal, human response to overwhelming circumstances, and it's an important step in getting past the stress and trauma of the event.

Any therapist will tell you that the only way out of emotional pain is through it. As Marcel Proust wrote, "We are healed of suffering only by experiencing it to the full."

EXPRESSING FEELINGS

Acknowledging what you are feeling makes you feel less victimized by the event and releases you from the emotion's stranglehold. This gives you back a sense of control. Expressing feelings can also help you connect with the social support and love you can get from others.

You only delay your ability to deal with crisis when you buy into old-fashioned views of how you should react—feeling that you're a sissy if you cry, or you're not feminine if you get angry, or it's not acceptable for you break down emotionally. Such reactions only injure you more. Author Jana Lee Frazier puts it this way: "To imply human beings shouldn't react to loss or abuse with debilitating emotional pain is like expecting the body not to bruise or bleed after a physical beating."

Instead of holding onto these outdated views, it helps to recognize that emotions serve a very important function. Anger signals that we're not getting what we want or that someone is hurting us, or that we feel helpless in the face of tragic events. Fear helps us take wise precautions in our lives, and grief opens us to compassion.

Dealing with feelings will help you perform better in your job as a broadcaster. You can't deal with the next crisis if you haven't processed the previous one. In workshops I often use the analogy of the oxygen masks on airplanes. The flight attendant always instructs us to put our masks on first so we can be of assistance to others. Unless we have dealt with our feelings, we have nothing to give any one else.

Only when emotions are held in do they become harmful. When feelings are not identified and processed, we end up with emotional debris that accumulates inside us. The buildup of this emotional debris is what can overwhelm us and make feelings seem out of control. If we identify and deal with feelings when they present themselves, it's like putting out small brush fires before they become an uncontrollable forest fire.

KNOWING WHEN AND HOW TO DEAL WITH FEELINGS

You may be squirming with great discomfort at this point at the idea of spilling your guts and weeping and crying or raging in anger. Fortunately, that's not always necessary. You have many choices as to how you can deal with feelings after a crisis. We'll look at several of them in this chapter.

One of the most important points I want you to remember is that *you* determine how and when you deal with your feelings. Dealing with feelings cannot be rushed, and it cannot be "one size fits all." You should go at your own pace and seek out a way to process your emotions that is right for you.

For example, you may find that right after a crisis event you want to delay dealing directly with your feelings and instead do something to channel your feelings into action. You might give blood or do charity work or, as Mike Walter did, give interviews to talk about the event (see chapter 9). Go at your own pace and don't let anyone push you into facing your feelings before you're ready.

WHAT CAN YOU EXPECT TO FEEL?

The first step in facing your feelings is to identify them. A common slogan is "You can't heal what you don't feel."

So what can you expect to feel? According to noted psychiatrist Elisabeth Kubler-Ross, the standard emotional stages of a loss or trauma are denial, anger, bargaining, depression, and acceptance. But don't think that these are neat little stages and that you go through them once and you're finished. The length and intensity of the trauma you've experienced usually dictates the length and intensity of the feelings. Two years after September 11, for example, 19 percent of rescue workers were still experiencing signs of Post-Traumatic Stress Disorder. That's double the rate in the general population.

You may also have a combination of feelings all at once, and you may cycle through them more than once. But the good news is that things change, tears run out, and no feelings last forever at the same intensity if they are processed in some way. In his book *Teach Only Love,* Gerald G. Jampolsky explains: "There is an end to pain. There is a point beyond which misery cannot go."

Before we explore ways to process feelings during times of crisis, let's look at some of those feelings more closely.

Denial

Responding to a crisis with denial is not necessarily a bad thing. In fact, I think it's incredibly helpful if a broadcaster:

- Uses black humor to avoid the pain of an event
- Denies an event was as bad as some say
- Blocks it out
- Avoids thinking about it
- Says, "I'll deal with this later"
- Compartmentalizes the trauma

I say these are helpful because they are ways to cope when you have to keep working and the trauma is too much to deal with at the time. These approaches are great to employ when you're dealing with a crisis at the moment it's happening. How many surgeons do you think allow their emotions to be felt when they're cutting open someone's skull or amputating a limb? Distancing themselves is the only way they can do the procedure. They must think of their technical skills only. The same is true for you during a crisis. But the key word is *during*.

While denial can be a great tool to use during the crisis event, it's a destructive tool later on. It stores away the trauma and can block your healing process. Caryn Summers explains: "Our most profound growth comes during our most painful times. . . . When denial stops, the process of healing begins. In the center of chaos lies the promise of clarity."

We are able to keep going during a crisis not just because we block out our emotions through denial, but also because of the adrenaline that is surging through our bodies. Soldiers, for example, can keep fighting even after they're wounded, because adrenaline blocks the pain. Adrenaline can also block negative feelings. One approach many people take when they feel even a hint of emotional pain is to keep themselves in a state of crisis so that the adrenaline will do just that. They may work long hours or add more activities into their lives—anything to keep busy and avoid facing their feelings. This may work in the short term, but it's no long-term solution.

Novelist P. D. James describes what can happen when feelings are avoided for too long. She says they can become "like dangerous and unpredictable beasts that lie curled in the pit of the subconscious." Many people turn to drugs or alcohol or work to avoid facing these beasts. But, when you suppress the intense feelings of a crisis for an extended time, you usually suppress your ability to experience good feelings as well. Your continuum of feelings becomes very short. You never feel anything intensely.

Suppressing feelings can affect your ability to do a good job on air as well. During a recent crisis in Maryland, I heard a reporter at the scene of a train crash say, "The bottom line is, two are dead." When I asked him why he delivered this news in such a cold manner, he said he didn't really think it was that tragic because only two people had been killed. He seemed to have suppressed his feelings for so long that he was not able to have any empathy for the family and friends who were listening to his report fearing their loved one had been killed.

According to *The Post-Traumatic Stress Disorder Sourcebook,* "Suppressing powerful emotions requires such effort that it takes a devastating toll on health." Many of the most common complaints that make people miss work, such as back pain, headaches, and stomach problems, are directly linked to stress. We know that stress can make us sick, and suppressing emotions creates stress.

Grief and Depression

A normal reaction to a crisis is to feel grief. On anniversaries of September 11, we all feel grief for what happened that day in 2001. One good thing that has come out of that event is that it has taken grieving out of the closet. People now openly grieve together. It wasn't too many years ago when Rep. Patricia Shroeder was rejected as a serious presidential candidate because she teared up in public. Now we see men crying at the site of Ground Zero. When two snipers terrorized the D.C. metropolitan area, one of the most powerful moments during those weeks was when Police Chief Charles Moose cried during a press briefing. It's no longer taboo to show grief or to shed tears in public.

Researchers have discovered that the tears we cry when we express grief are chemically different than the tears cried when we do something like cut up an onion. The tears produced by the onion are mostly salt water. Tears we cry from sadness appear to produce proteins that are precursors to the production of natural antidepressant hormones in our bodies. Crying literally helps us recover from trauma. Leonard Felder puts it this way: "Just like heavy rainfall clears the air and is followed by the sweet sounds of birds singing, so does a good cry bathe our insides with a healing release." In short, crying is good for us, and it's a natural way to express grief.

Grief may not surface for you until several weeks after a trauma. Don't assume that you'll go home the first night of covering a crisis and cry your eyes out. As you'll read in the Profiles of Crisis, most of the reporters didn't deal with their feelings right away. You may find that weeks after the crisis you begin having trouble sleeping or start feeling down. You may get exhausted easily or experience mood swings. These can all be signals that you need to deal with your grief.

If you begin to have more severe symptoms that last more than two weeks, you may be suffering from depression. This is very different from feeling sad or down or blue. Depression is more severe than that, and it can be incapacitating. If you have two or more of these symptoms for two weeks, it is a good idea to seek professional help:

- Trouble sleeping or sleeping too much
- A feeling of hopelessness
- A lack of joy in life—nothing makes you feel good
- Loss or increase of body weight
- A feeling of restlessness
- A feeling of listlessness
- Suicidal thoughts (this requires immediate attention)

Suicides in our country outnumber homicides five to three. Nearly thirty thousand people commit suicide every year, and 70 percent of those are white males. Take any suicidal thoughts very seriously.

People who are depressed are three to five times more likely to die prematurely, so it's important to take depression seriously as well. The methods that you'll read about later in this chapter will help you deal with grief and recognize depression.

Fear

Another common feeling following a crisis is fear. This may be even harder to admit than grief. Fear is a difficult feeling for everyone. It's not coincidental that the phrase "Be not afraid" is repeated about three hundred times in the Bible. Fear has always haunted us as humans.

Crisis makes us come face to face with the fact that the future is uncertain. Life can change in an instant. Reporters see this every day. Desirée Berenguer talks about what it's like to face these daily events that illicit fear (see chapter 14). Even Walter Cronkite talks about the fears he felt in World War II in his interview with Mike Freedman (see chapter 17).

But, while fear makes us face our mortality, it also helps us appreciate life. Almost all the reporters who wrote profiles for this book talk about how they learned to value life more because of the crises they covered. When we see that the future is uncertain, we can also see the unlimited possibilities for things to be better in our lives and in the world.

Anger

Anger is the "fight" part of the fight or flight response. It's usually triggered because we are not getting something we want or need or because we feel helpless in the face of events. These are all legitimate feelings, but anger is one of the most destructive responses when it is expressed in the wrong way. Road rage and violence of any kind are linked to anger. But you don't have to go out and kill people for anger to be a problem. Suppressed anger can be especially toxic and dangerous.

Suppressing anger makes us more likely to have heart disease and many other illnesses. In fact, harboring hostility is a more powerful predictor of heart disease than high cholesterol, smoking, or obesity, according to a three-year study conducted by Dr. Raymond Niaura at Brown Medical Center. This is because anger keeps us in a stress response. If we suppress anger, it's like stepping on the accelerator and the brake in our car at the same time. This stress literally pumps up the adrenaline in our bodies without allowing any release.

Anger is an emotion that begs to be expressed. It often requires action to really feel the anger and burn off the adrenaline. Sports are a great way to do this. Going to the gym or doing kickboxing or karate can work wonders with expressing anger. Remember that anger is not an unhealthy feeling. It only becomes unhealthy when it's suppressed or acted out in harmful ways.

IDENTIFYING FEELINGS

Many times, if you're like me, you may have no idea what you're feeling at a given moment. After a crisis, there are several ways to help get in touch with your feelings and identify them.

As you do so, however, it's important to monitor your reactions. If at any point the process becomes overwhelming, do not attempt to continue alone. Ask a friend or loved one to help you, or seek out a mental health professional. (Getting professional help is mandatory if you begin to feel like hurting yourself or someone else.)

It's also important to set aside a time to deal with feelings. None of us want to go through the day dragging around the heavy emotions that the process can bring to the surface. I suggest you make a daily appointment with yourself to sit down and work on processing some of your emotions. At the end of this time, imagine you're putting all the feelings in a drawer. If you've been writing, you can literally put your journal or notes in a drawer. Close the drawer, take a deep breath, and walk away from the feelings for a while. Know that you have set a time to come back and deal with them the next day. Identifying and healing strong feelings is something you only need to visit during your day. You don't have to live there. Remember that Focusing on Fun is part of the process of taking care of yourself. Consult chapter 8 and let joy and fun be a big part of every day while you're dealing with feelings.

So how do you start to process your emotions? Simply naming what you're feeling is a good first step. Here are some words that might help you in this process. Make several copies of this page and fill in the first blank in the sentence below on one of the copies. Next, circle as many words as you want to that could go in the second blank (you can add your own words as well).

Naming Feelings

When I see footage of or think about ________________,
I feel ________________.

Sad	Terrified	Angry
Defeated	Anxious	Enraged
Hopeless	Scared	Hostile
Helpless	Frightened	Irritated
Dead Inside	Threatened	Aggressive
Vulnerable	Endangered	Cynical
Confused	Insecure	Hateful
Alienated	Agitated	Outraged
Isolated	Edgy	Explosive
Victimized	Horrified	Vengeful
Violated	Panicky	Out of Control

Date______________

Labeling feelings is a good way to get in touch with them. You will probably find that if you do this exercise every day you'll circle different words at different times. Feelings change.

If you want to take this exercise a step further, rate each feeling that you've circled with a number between 1 and 10. A 10 should be the strongest you could possibly feel this feeling. On September 11, for example, I would say my sadness was a 10. A month later, though, it may have dropped to an 8. Notice the gradations of feelings and how they are constantly changing. You can also use this rating system during an actual crisis event. In chapter 8 you'll read about how to use this rating system to calm yourself down during a crisis.

Once you've gone through the steps above, allow yourself to sit calmly with the feelings you're having. Don't dismiss them or try to numb yourself to them. You may find yourself crying or feeling angry or frightened. Any feeling is okay. As you're experiencing this feeling, recognize that it is not overwhelming you. (If it feels like it is, you can share it with someone, or stop the process.) If you label what you're feeling and give yourself time to feel, some of the force of the emotion will usually fade away. You will also see that experiencing the feeling will not kill you. It's just a feeling.

The six-year-old daughter of a friend of mine summed up identifying feelings better that I ever could. When her mom was upset, she told her, "If you let go of your feelings, they'll let go of you." That really says it best.

As you identify your feelings, there are two other questions that may help you deal with them. First, ask yourself, "What do I need right now?" When my sadness was so intense on September 11, I found I needed to look for what was still normal in the world. Melinda Murphy needed to connect with her husband (see chapter 10). Sherry Stalley needed to watch her children sleep (see chapter 11). When negative emotions are strong, find out what you need to help yourself feel better.

The next question that is helpful to ask is "What am I feeling in my body right now?" Try to identify how you feel physically. You may find that your stomach is in a knot or your head is hurting. Your heart may be racing or your mouth may be dry. Feelings are often held in our bodies. They need to be released. Once you've identified what you're feeling physically, you can take steps to release some of the additional stress that has come into your body. Holding the stress in your body can result in illness or discomfort.

WRITE ABOUT IT

Once you've identified your feelings, or to help with that process, you might want to consider writing as an outlet. Writing your feelings down helps in many ways. Have you ever tried to remember what you need at the grocery store? If you don't write it down, you will be thinking about it off and on for most of the day until you get to the store. If you write out a grocery list, you can forget it until you pull the list out at the store. The same is true for feelings. When you write them down, you can forget them until you pull out your journal or notebook later.

Writing can also organize your thoughts. Without this organization it often seems like feelings are swirling around in your head. It's as if you're rehearsing the feelings all the time. If you write about it, you can calm your thoughts and release the feelings.

Just like you learned how to write for broadcast, you may need to learn how to write to process your feelings. First, this is not like writing a television or radio story. It's not the Who, What, When, Where approach. In fact, it's the opposite of this. Instead of writing about what happened, you want to write how you *felt* about what happened. That's the key to this type of writing.

And some studies show that the other helpful thing is to write about how you're dealing with what you're feeling. A University of Iowa study showed that recording events was not as helpful as writing about emotions and then writing how the emotions are being dealt with. For example, after the D.C. sniper incident, I might have written, "*I'm feeling lots of fear that another sniper might come along and start killing people, but I'm reminding myself of the small odds of that happening. That makes me feel calmer.*" This is the type of writing that has proven most beneficial.

So let's look at the way you might want to use writing to process your feelings. First, it doesn't matter whether you buy an expensive journal to write in or you use one of your reporter's notebooks. I have friends who like to do all their writing on their computer. Choose the way to write that fits you and your lifestyle best.

It is a good idea to set aside a time to write when you won't be interrupted. Find a quiet place where you can spend at least ten minutes alone. Don't worry about spelling or grammar when you write. This is only for you (unless you choose to share it with someone).

It's usually a good idea to begin with some of the feelings that are the easiest to deal with. Begin with the five-pound weight and not the hundred-pound one. For example, suppose you've just covered a shooting that reminds you of the loss of a loved one to a gunman. Rather than jumping right into your feelings of grief, you might focus first on your anger that your cameraman didn't get the images you wanted at the shooting scene. Work your way up to the most intense feelings that you have, and take your time. If you get overly distraught, always stop and call a friend or loved one and talk about your feelings.

Remember to put your writing away in a drawer when you're finished so that you have a sense of closure. Sometimes you may feel tired or sad after writing, but this usually shifts rather rapidly. If it doesn't, try talking to someone about it.

Here are some phrases that might help you get started with your writing. Simply complete these sentences:

1. The idea of writing about my feelings feels . . .
2. I'm most afraid of . . .
3. I wouldn't want anyone to know I feel . . .
4. My life feels . . .
5. I'm angry about . . .
6. I'm sad that . . .

Try these phrases to see if they help you get started with your writing.

Another way to begin is described by Martha Beck in her book *Finding Your Own North Star* (see Suggested Readings). She suggests you start with a blank page and first write, "This is what happened." Fill in the space below with what happened. Next write, "This is how I feel about it." Again write about this. Now take out a fresh piece of paper and write, "This is what REALLY happened." Then write some more. Next write, "This is what I REALLY feel about it." She explains, "Dig deeper into your gut reactions, *taking especial care to describe any feelings you think may be 'inappropriate'*" (italics hers). This can be a very effective exercise. Her caveat for this exercise is the same recommendation I've made throughout this chapter, however: *"Don't do this exercise until you've established at least one supportive relationship, if possible, with a qualified therapist"* (italics hers).

TALK ABOUT IT

Writing is an excellent way to get in touch with your feelings and process them. Sharing your feelings with another person can also be very healing. In fact, a study done at Stanford University in 1980 found that breast cancer patients with metastasized cancers who attended a weekly support group in which they talked about their feelings lived an average of eighteen months longer than those who did not (twice as long). They also suffered less pain. The support group provided several things that were important. First, the women felt listened to and cared about. Next, they felt they had some control, because going to the group was something they could do. And, lastly, they felt their feelings were not crazy and that what they were feeling was explainable.

(Note that occasionally support groups do not work well. Mike Walter talks about group experiences that did not work for him [see chapter 9]. Sometimes, hearing others share their pain only makes yours worse, as it did for Mike. You may feel you're not reacting the same way as the other group members and that that's not okay. Always go with what feels right to you and don't stay in a group that you don't find helpful.)

Rudy Giuliani found that talking really helped when he was diagnosed with prostate cancer. He writes, "The best thing that helps when you're dealing with fear, stress, or things that confuse you is to talk it out with people. Talking to and relying on other people . . . are enormously helpful."

One of the ways that talking with others is helpful is that you realize that other people are having or have had the same feelings you're having. You may feel that because you saw or reported on things during a crisis that you're somehow different from everyone else on the planet. You may that feel no one understands what this is like but you. Give them a chance. Human feelings are human feelings and have been felt by humans for thousands of years. We can all share feelings even if our experiences are different.

Social support is so important to us as living beings. Many animals have the same needs. In one study, for example, reported by Robert Sapolsky (see Suggested Readings), infant primates were put through a stressful situation. If they were in a room full of other primates that were friends, the stress response lessened. If they were with strangers, it increased. Social support helps control stress. This has been seen with Holocaust survivors as well. *The Post-Traumatic Stress Disorder Sourcebook* reports that "low disclosers were found to be the least healthy" over time. Talking about it can lessen the stress and thus keep us healthier.

The first step in talking about your feelings is to protect yourself by finding a safe place and a safe person. When we don't have a safe place to talk, we may continue to repress feelings. This is our subconscious way of protecting ourselves from being hurt, by talking in the wrong place or to the wrong person. None of us wants to be ridiculed or rejected.

Once you've begun talking, you may find you have to tell your story over and over to the same or different people. Sharing your feelings once doesn't usually lessen all the pain. Each time someone shows compassion and caring for you, the pain will be reduced. It's like carrying around a one-hundred-pound weight. That would be pretty exhausting, but if you give ten pounds of it to ten people it's manageable for everyone.

There are some people to whom you might *not* want to go for compassion and understanding. Ironically, the first to avoid might be others who have been through the same thing. This may seem like strange advice, but they may not have much compassion to give you because their own feelings may be so intense. A friend of mine who lost many colleagues on September 11 found that sharing her feelings at work was not a good idea. Everyone was dealing with intense feelings there, and she was not able to get the support she needed.

You may also find that talking to your partner, spouse, or friends is not too helpful either, especially during a crisis. Jim Farley, the vice president of news and programming for WTOP News Radio in D.C., shared with me that loved ones are often so worried about reporters covering crises that when the reporters open up to these people about their fears or sadness the only response they get is that they should quit their jobs. This is certainly understandable but not at all helpful. I would hope that there are many bosses out there who are as willing as Farley is to make a box of tissues available in their offices and to provide a safe place for staff members to express their feelings. (He tells me that tissues have been a fixture in his office since he first got into broadcast news management in 1976.)

It's also important not to ask friends and family to be your counselor or therapist. Since it is helpful to retell your story and talk about your feelings many times, this can be a burden on any relationship. A trained mental health professional is a good choice not just because he or she has lots of training—although this can be enormously helpful—but also for the actual fact that you pay him or her. Because mental health experts are paid they can devote hours to listening. They can also provide a safe place where you will not be interrupted while you process your feelings. Talking to a professional is a way to purchase what you really need when you're processing trauma. He or she will be totally committed to being present for you and really listening, with no other job than to help you feel better.

Going to a mental health professional does not mean that you're crazy. There are many types of therapy and many uses for therapy. If you're considering this option, I suggest you put your best reporter skills to work and interview many therapists. Talk on the phone to as many as you need to, until you feel you've found someone you can be comfortable with and talk to easily. Then see this person for a month on a trial basis. If you're not feeling comfortable with this person after a month (or sooner if you get a strong negative reaction), end the relationship and begin calling other therapists. Therapists are as different as any other group of professionals. The right one will be out there for you.

MOVING PAST THE FEELINGS

The goal in identifying your feelings is to understand what this trauma means to your life so that you can integrate it and become stronger and more resilient because of it. Resiliency is the key to healing. Bad experiences are always going to happen, and if you're a human being you're going to have lots of feelings about them. But feelings don't have to make us victims. They can motivate us to make life better for ourselves and others. Identifying feelings and processing them gives us strength. As author Rachel Naomi Remen says, "The process of wounding awakens us to our strength." Another way to look at this was written by Aung San Suu Kyi, the 1991 Nobel Peace Prize winner and democratic leader of Burma, who has spent many years under house arrest. She writes, "It is from hardship rather than ease that we gather wisdom." Feelings are a gift that opens the door to that wisdom.

The Bottom Line of Identifying Your Feelings

1. It is normal to have strong feelings after a catastrophic event.
2. You are the only one who can decide when and how to deal with your feelings
3. There's a time for denial and blocking, but in the long run they are harmful.
4. You may feel anxiety, fear, sadness, grief, anger, or depression.
5. Always seek help from a mental health professional if you feel overwhelmed or have any suicidal thoughts.
6. Feelings may be felt in the body.
7. Writing about your feelings will help you organize them and process them.
8. Talking in a safe place with someone who can offer you compassion lessens the stress.
9. Wisdom and resiliency come from processing feelings.

CHAPTER 7

E = Eat Well

"To keep a lamp burning you have to keep putting oil in it."

Unknown

Did you know that what you eat affects how you report a story? Well, this is the case. Good nutrition gives us the energy to keep going and the mental alertness to make wise decisions. It also gives our immune systems what they need to stay healthy.

Such benefits are never more important than in a crisis. But it's often at this very time that we turn to foods that are loaded with sugar, caffeine, and alcohol, which actually make the crisis worse. Or perhaps we quit eating because the stress is affecting our digestion and we're just not hungry. This was the case for Sherry Stalley during her coverage of Flight 93 on September 11 (see chapter 11). But poor nutritional choices can actually make the crisis seem worse, because of their effect on our blood sugar levels. Good nutrition should be a top concern for you when you cover any stressful event.

It's difficult to analyze everything you're eating when you're in the midst of a crisis. But, if you understand a little about what food does to your body, you're more likely to make good choices under stress.

I like to compare eating to maintaining a car. A car needs two things to run well. It needs gasoline, and it needs a good mechanic. Our body needs food to run well, and food supplies both the gasoline and the mechanic. The carbohydrates we eat are the gasoline. They supply us with glucose, or blood sugar, which gives our muscles and brains what they need to operate. Protein is like the mechanic, because it repairs, renews, and rebuilds our cells. It's only by eating a combination of protein and carbs that we maintain a healthy, well-functioning body. Fat is the third nutrient needed for good health, but in a time of crisis it's not necessary to concentrate on fat intake. In daily life when you're not in a crisis situation, it helps to aim for low-fat foods for weight control.

If you learn a little about how protein and carbs work within our bodies, you'll better understand their importance.

CARBOHYDRATES

We've all been told to eat our vegetables and strive for five fruits a day, but you may wonder why this is important. Fruits and vegetables are a strong source of carbohydrates, which supply the blood sugar that fuels every cell in your body. They are your main source of energy. A simple way to remember what is classified as a carbohydrate is to think of carbs as anything that grows out of the earth: fruits and vegetables; sugar; grains; potatoes; legumes such as peas, beans, and peanuts; or the products made from these plants.

As soon as we eat a carbohydrate it is converted to glucose. Insulin, which is secreted by the pancreas, helps escort the glucose into our cells so that it can provide energy. Some of the energy is used immediately, and some of it is stored as glycogen in the muscles and liver to provide energy later. So, whenever you eat a slice of whole wheat bread, your body gets the gasoline it needs to operate.

Problems arise, however, when we don't eat the best carbohydrates. There's a significant difference between eating that whole wheat bread and eating a Krispy Kreme doughnut. The difference lies in how fast the glucose enters our bloodstream. A doughnut is a *simple carbohydrate.* Simple carbs, like refined sugar, white flour, or white rice, deliver glucose to the bloodstream very fast, perhaps within only fifteen minutes. The white flour and refined sugar in a doughnut create a fast one-two punch to your glucose level. This may feel great while you're eating the doughnut and shortly afterwards, but in twenty or thirty minutes you will have less energy than you did before you ate. Simple carbohydrates give you a sudden spike of energy and then a quick crash (see figure 7-1).

If you think about primitive man, you'll get a clue as to why this happens. In order to get the amount of sugar that's in a single doughnut, primitive man had to eat a huge amount of leaves and berries. The body is still designed to handle this type of eating. When this pure sugar comes into our system now, our insulin level goes way up to escort all the food the body thinks it's getting into the cells. Since

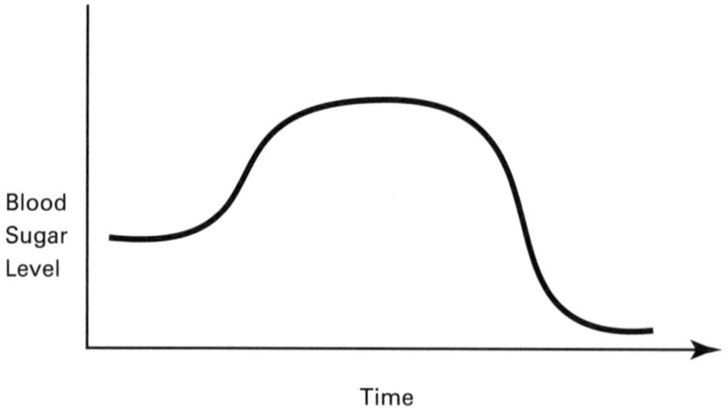

Figure 7-1 Eating refined or simple sugar may cause your blood sugar level to drop even lower than it was before you consumed the sugar. (Adapted from *Dr. Dean Ornish's Program for Reversing Heart Disease.*)

there's not much food coming in, too much sugar is taken out of our bloodstream, and we crash.

However, some carbohydrates don't cause this response, and they're called *complex carbohydrates.* These types of carbs take about two hours to digest, so they don't cause the spike in blood sugar. They provide good, lasting energy. Here's a list of complex carbohydrates:

Complex Carbohydrates

Whole grain breads and cereals (the word "whole" should precede the type of grain in the nutritional ingredients)

Brown rice

Vegetables

Legumes (beans, peas, and peanuts)

These foods give us the most constant, steady supply of energy we can get. They prevent the sudden spike in blood sugar that can create more stress on our bodies. They also increase the secretion of serotonin in the brain, the feel-good chemical that improves our mood and calms us down. All of this sounds pretty good when you're working in a crisis situation.

Sometimes we shy away from carbohydrates of any kind, thinking that they will cause us to gain weight. But cutting back on carbs means that your glucose level drops, and with it your energy.

When you're working during a crisis, you should not worry about gaining weight, for two reasons. First, you will be burning more calories during this stressful time. If you were simply sitting still during a stressful time you would be burning more calories, but you will no doubt be running nonstop when you're at work. Don't think that you need to skimp on complex carbohydrates.

The other reason you shouldn't worry about gaining weight is that it takes 3,500 *extra* calories to gain just a pound. Weight gain usually comes from overeating for weeks at a time. During the few days at the peak of any crisis, you can eat a good deal more than usual and not worry about weight gain, because of the added stress and exercise you'll be getting. But always stick to complex carbohydrates and protein. Stay away from simple carbs if you can. For example, a 14 oz. slice of the Cheesecake Factory's carrot cake will give you almost half of the 3,500 calories you need to gain that pound.

PROTEIN

When we look at our bodies, what we're seeing is protein. Excluding water, protein makes up 50 percent of our body weight. Every single cell in our body contains some protein. The amino acids in protein are the main building materials of our bodies. They repair and renew cells. They also provide the basis for our immune systems by making antibodies that can fight disease. They serve in the creation of hormones as well. In short, without protein, we wouldn't be here.

Just as there are two kinds of carbohydrates, simple and complex, there are two types of protein, incomplete and complete.

Complete proteins contain all the amino acids our bodies need to function optimally. An easy way to remember which proteins are complete is that complete proteins come from animals that walk on the earth or swim in the water or are the byproducts of these animals. Here's a list of the best complete proteins, in order of their value:

1. Eggs
2. Milk (including yogurt)
3. Cheese (including cottage cheese)
4. Lean meat, fish, poultry.

On the other hand, *incomplete proteins* provide only some of the amino acids our bodies require. These types of proteins are effective only if combined with other foods that provide the missing amino acids. For instance, if you eat peanut butter on whole wheat bread or crackers, it becomes a high-quality protein. The same is true for beans and rice; one or the other by itself is incomplete, but by combining them they become complete proteins.

In a time of crisis, it's important to eat as many complete proteins as you can, since you will not have time to think of combining foods in a special way. Looking at the list of complete proteins will give you an idea of what will give you the gold standard of protein at a time when you most need it.

One of the problems with protein, whether it's complete or incomplete, is that the body does not store it well. The amino acids in protein flow into our bodies and can be used for a few hours. After that any that haven't been used are converted into carbohydrates for energy, or fat. So it's important to keep a supply of protein coming into the body all day long.

EATING ENOUGH

For many people one of the biggest difficulties in time of stress is eating enough. Because the stress response shuts down the digestive system, food often has no appeal. But, if you consider the role carbohydrates and protein play in bodily health, you should realize that food is important even when your mind tells you you're not hungry.

Many of my clients say that they don't feel like eating first thing in the morning, so they only have a couple of cups of coffee. Once they get to the station, food doesn't enter their minds or mouths until after their shift is over. By that time they have essentially been fasting all day.

When we don't eat, fuel must come from inside our bodies, and it's not always from our fat stores as we would like. Muscle tissue suffers, along with other tissues. As we've seen, the body must have energy to keep going, so it will forfeit the necessary protein functions to convert protein to fuel. When this happens over a period of time, our bodies and our immune systems will suffer. During a time of crisis, you don't want to find yourself sick, and this can happen if the body doesn't get the nutrition it needs.

When you don't feel like eating, think of food as a drug that has been prescribed every three or four hours. Eat something that contains carbs and protein just to keep your body working well. You may not enjoy eating it as much as you would if you were having a relaxing meal at a nice restaurant, but you must get the nutrients into your body.

EATING EVERYTHING IN SIGHT

Some people have the opposite problem from eating too little when they're stressed. They want to eat constantly. As we saw in chapter 1, stress increases the glucocordicoid levels in the body, and this is linked to a craving for fats and carbs. This is a result of the intense need for energy during a crisis; the body wants to replenish glucose levels after a stressful day. Remember that simple carbs provide glucose in its purest form, which may be why we want to grab the ice cream or Oreos in the evening and not the pot roast.

If you find that you want to eat when you're stressed, it's important to look at what you're eating. It's easy to consume way too many carbs under stress. Stopping at Starbuck's, for example, for a quick snack can result in almost a day's worth of calories. Their muffins range from 400 to 500 calories each. Combine this with a coconut crème frappacino at 870 calories, and you have consumed most of the calories you need for the whole day. Plus, the refined sugar content will send your energy soaring only to plummet a short while later (see figure 7-1). Eating more than 1,000 calories at any single time will make you sleepy, irritable, and inattentive afterward.

If you feel the need to eat while under stress, look for complex carbohydrates like vegetables or whole grains, or some protein, which will satisfy your craving without increasing your stress and your calories.

GRAZING

It takes a full twenty-four hours for some foods to be fully digested, but food is available for energy in its purest form for only a few hours. Because of this fact, *grazing*, or eating a higher number of smaller meals a day, is the best way to keep your energy level up. Think of how monkeys and apes eat. They graze on food all day long, grabbing a banana here and a few ants there. This is probably the way our systems expect food to come into our bodies. Also, our brain cells cannot store glucose like other cells in our bodies. Therefore, they need to have food coming in at a steady rate all day to keep operating effectively.

Obviously, if you're going to eat four or five times a day, you want what you eat to be healthy and low in fat and calories. As we've seen, the combination of carbohydrates and protein provides the best balance for the body and releases a steady supply of glucose to keep energy available. But food must continue to come into the body every three or four hours for everything to remain in balance.

When you practice grazing, your eating pattern during the day may look like this:

1. Breakfast
2. Midmorning snack
3. Lunch
4. Midafternoon snack
5. Dinner

Each of these meals should contain approximately two-thirds carbohydrates and one-third protein. Remember the simple way to distinguish carbs from protein: carbs grow out of the earth, and protein comes from animals that walk on the earth or swim in the water. You need some of both of these every three or four hours during the day. Crisis Survival Kit 4 at the end of the book gives you a list of healthy foods you can easily carry in your bag or keep in your desk to make grazing easy.

NUTRITIONAL LANDMINES

Before we look at healthy ways to fulfill your grazing needs, let's consider some of the common pitfalls we face when we're hungry and stressed. One of the first places we may find ourselves literally driven toward is a fast-food restaurant. When time is short and we must eat something, fast food often seems to be the only choice. It's not unusual that we have this response, since McDonald's spends more advertising dollars in twelve hours than the National Cancer Institute spends in an entire year on its campaign to promote the healthy alternative of fruits and vegetables.

But stopping at a fast-food restaurant doesn't have to mean dietary disaster. You can now get baked potatoes, salads, fruit and yogurt, and other choices that are healthier than the Big Macs and Whoppers, fries, and milkshakes or Biggie soft drinks. A quick stop at a grocery store, however, is always a better choice than fast food, and it will only take a few minutes longer.

The other nutritional landmine lurks in every building—the vending machine. Here the healthy choices are limited. Some machines may offer fresh fruit or nuts, but most of what you find will be simple carbohydrates like cookies, pretzels, and candy. Even the juice sold in soda machines is really soda disguised by its uncarbonated form. Juices labeled "100 percent fruit juice" may contain much more sugar than you thought. One small box of cranberry/raspberry drink, for example, may have as much as 38 grams of sugar. Four grams of sugar is equal to one teaspoon of table sugar, so this boxed drink would give you over nine teaspoons of pure sugar. That's more than a doughnut. This drink would hurt your stress level, not help it.

Vending machines are not the only lurking threat in most buildings. A client who works in a television bureau where the deadline pressures are especially intense told me this story. Every day at deadline time, the clerk from the coffee shop in the building comes through the newsroom selling candy bars. They nicknamed him "Chocolate Bob," and my client said when she looks up from her computer and sees him, it's as if he has a halo over his head. He can overcharge for the candy. No one cares. They just need that sugar fix. And, when he doesn't show up for some reason, everyone suffers, because they're so used to that spike of energy the chocolate gives them.

Another time when unhealthy food may be in abundance during a crisis is if catered food is brought in or a craft services table is set up. Because most simple carbs don't need to be refrigerated, they are really easy foods to grab on the run and to have available all day. These tables are usually full of cookies, cakes, doughnuts, pastries, bagels, bags of chips, and packaged crackers made with white flour and topped with cheese or peanut butter. There may be fresh fruit available, and that's always the best choice. If there's a cooler with yogurt, cheese, or milk in it, those will provide excellent protein choices. But you will most likely see only simple carbs provided.

Nutritional landmines are not limited to places away from home. Your kitchen cabinets may provide all the cookies, sweet cereals, candy, pastries, and chips you need. Sometimes there's no escaping the wrong foods. Here's what Page Hopkins of FOX News Channel told me about her experience when she first became a reporter in a small market:

> I ate very poorly when I started reporting. I was a junkaholic. I ate lots of fast food, and as a result I was having stomach and intestinal problems. My hair even started falling out. Every morning I would eat sugared cereal with milk and find myself tired for a couple of hours afterwards. I was feeling cloudy and foggy because of the sugar, and what would I do to cure it? I would have a Snickers and a diet Coke. The worst possible thing I could eat. And I wondered why my hair was falling out! I realized how important food was when I started looking at tapes of myself. I could tell the days when I wasn't eating well. I could see subtle things like a lack of focus when I looked at the camera or a slower rate.

Page became so attuned to her nutritional needs that she began bringing a cooler to work with her packed with healthy foods. She found that, without the proper foods in her body, her work suffered on a daily basis.

HEALTHY EATING IN A CRISIS

So what can you do when all that seems to be available are fast-food places, vending machines, and unhealthy food choices? The first step is to stock your refrigerator and cabinets with the best foods you can. If you have good foods to choose from, you will be prepared when a crisis strikes.

Here's a list of some foods to buy when you shop:

Healthy Foods to Buy

Low-fat milk

Low-fat cottage cheese

Low-fat yogurt

String cheese

Eggs or egg substitutes such as Egg Beaters

Lean meat

Poultry

Fish (both canned and fresh or frozen)

Canned meat or chicken

Soups with meat, chicken, or beans

Fresh or frozen vegetables

Fresh or frozen fruit

Nuts (lowest in fat are almonds, pistachios, walnuts, and hazelnuts—limit yourself to a handful a day)

Peanut Butter

Whole grain breads

Whole grain cereals

Brown rice

Whole grain crackers

Stocking up on these foods will give you healthy choices at home, and they will also give you some options for food to take to work each day even when you're not covering a crisis. On your first day of work each week, for example, bring in five or ten pieces of fruit to keep in your desk. With fruit on hand you can always grab something healthy for a snack. If you also keep a jar of peanut butter or some nuts in your desk, you'll have a healthy snack of carbs and protein.

It's also a good idea to have an "emergency food pack" that you can grab when you know you will be covering an event like an approaching storm or the D.C. sniper. Neal Augenstein found it hard to get food once he was at police headquarters (see chapter 12). If he'd had his emergency food pack ready to go, he could have grabbed it on his way out the door. Things in this pack should be nonperishable, easy to open, and healthy. Here are some choices that meet these requirements:

Emergency Food Pack

Peanut butter

Whole grain crackers or breadsticks

Whole grain cereal

Small cans of tuna, salmon, or chicken with pop-tops

Protein bars (look for bars with less than 15 grams of sugar and at least 15 grams of protein)

Nuts or trail mix

Dried fruit and raisins

Fruit juice in drink boxes or cans

V8 juice

Liquid meals like Slim-Fast

Water

Plastic fork, knife, and spoon and zip-top plastic sandwich bags for storage

You can always be prepared with food that will give you good nutrition under the worst of circumstances. These food items can be stored in a bag for months without spoiling. A quick stop at a grocery store can make your food options even better; add some milk, fresh fruit, string cheese, yogurt, lunchmeat, whole grain bread, etc. But you can make it through several days with your emergency food pack alone.

When you do get to stop at a restaurant when you're covering a crisis, think of more than just that meal. Order extra food to go so that you can divide it up for a snack and another meal. Pick up fruit whenever you see it for sale at a convenience store, a grocery store, or on the street so that you can keep it in your bag for later. Working a crisis means not knowing when your next mealtime will come, so plan ahead.

Breakfast is one of the most important meals whether you're covering a crisis or not. Think what your body must need after you haven't eaten for twelve hours. It's not called "break fast" for nothing. Many of my clients say that this is the one meal they can't stomach. I say, get over it. Find something you can eat no matter what time you have to get up. One reason that breakfast is tough for some people is that they think only within the box—the cereal box. Here are some choices that may seem nontraditional, but your body really won't care. It simply wants the protein and carbs it needs to get moving.

Nontraditional Breakfast Ideas

Tomato juice with a handful of nuts or two cheese sticks

Trail mix with an apple or cottage cheese

Peanut butter and jelly sandwich with milk

Yogurt and fruit

Egg salad sandwich

Hard-boiled egg and V8 juice

Whole wheat bagel with low-fat cream cheese or lox

Smoothie with protein powder

Protein bar

There are many other things you can eat for breakfast. Don't be afraid to experiment. One client told me he often eats leftover spaghetti with meat sauce or leftover chicken with potato salad. Whatever works.

Because stress can cause some digestive problems, one time when you don't want to eat is within two hours of going to bed. Too much food late at night will affect your ability to sleep well and can result in gastric reflux. See chapter 5 for more information on food and sleep.

If you want some help designing an eating program that works best for you, you can call the American Dietetic Association at 800-366-1655 to get a referral for a registered dietitian in your area. Anyone can call himself a nutritionist. If you decide to work with someone, make certain he or she has a legitimate diploma as a registered dietitian (R.D.), or an M.S. or Ph.D. in nutrition studies. He or she will understand the extra stress of covering a crisis and know more information to give you about eating well.

WATER

Water is the single most important nutrient in our bodies. If what we see on the outside when we look at ourselves is protein, what's on the inside is water. It's almost as if water invented the human body so it could move itself around.

All our bodily functions occur in the medium of water. Water regulates our body temperature, plays a vital role in digestion and waste removal, cleanses our blood, carries nutrients and oxygen to our cells, cushions our joints, and helps nerve impulse connection. The electrical and chemical signals in the brain move through fluid.

But the sad fact is that most of us pay more attention to watering our lawns and houseplants than hydrating our bodies—despite the fact that we are constantly losing water. We routinely lose ten cups a day by sweating, breathing, and eliminating wastes.

This water loss can be significant. Even a small amount of dehydration (around 1 percent of body weight) can affect our physiological state and our performance. Susan Kleiner, a registered dietician who has written a book on nutrition, says, "By the time you get thirsty, your body is already about 2% low on water. Losing 1–2% of your body weight in fluid can lower your physical and mental performance up to 20%."

Our blood volume is the first thing to suffer when we become dehydrated. This means that the oxygen in the blood will not reach our brains and muscles as effectively. Glucose must have oxygen to burn, so low blood volume means low energy.

Dehydration causes us to feel fatigued. In fact, that slump most of us get in the late afternoon is often caused by dehydration. For many of my clients, a big glass of water is as energizing as a candy bar.

During a crisis we can't depend on feeling thirsty to know when to drink water. Stress may override our bodies' warning system. To get the water we need in our bodies, we have to literally force fluids all day long. Water stays in our stomachs about fifteen minutes. For it to help hydrate our tissues and blood, we have to process it through our stomachs so that it can go out into the body. Therefore we need to keep water coming in all day long.

Does it have to be water? Many fluids will work, but they must be decaffeinated, unsweetened, and nonalcoholic. Other choices include:

- Low-fat milk, which is 90 percent water
- Yogurt, which is mostly water

- Herbal teas and decaffeinated coffee or tea
- Fruit juices (preferably 100 percent juice diluted 50/50 with water)
- Fruits, which vary from 50 to 75 percent water.

Some of these have drawbacks, however. Dairy products may cause your body to produce excessive mucous, which is a problem if you're on air. And it's hard to measure how much water is in fruit and yogurt. Flavored waters also can be a good choice, provided they do not contain sugar. Seltzer water and club soda may contain sodium, so check the label before consuming large quantities of either.

Looking at the choices, just plain water seems the best. Water is cheap and readily available, so there are few excuses not to drink it. To save money and ensure that you always have fresh-tasting water when you're covering crisis events, buy a bottle with a built-in filter from a company such as Brita. This way you can fill it up at any faucet.

How Much Water Should You Drink?

Calculating how much water to drink is simple if you use the ratio devised by the International Sportsmedicine Institute. They work with Olympic athletes and have done research on how much water we need. Their formula has been used for over two decades. On a day when you're not exercising aerobically, you need half an ounce of water per pound of body weight. So just divide your weight in half, and that's how many ounces of water you need daily. On days when you exercise aerobically, you need two-thirds of an ounce per pound of body weight. (A tall bottle of water like Evian is around 50 ounces.)

Your reaction at this point may be like most of my clients: good grief, if I'm out in the field covering an event, I'll spend all day looking for bathrooms! This may be true during the first week, but your body will quickly adjust to the increase of water. You might want to increase your water intake now so that, when a crisis strikes, you will not have a problem.

Some people fear that they could drink too much water and become ill. There is a condition called hyponatremia, which is a sodium imbalance brought on by excess fluid consumption during athletic events such as marathons. In this situation a person is sweating out large amounts of sodium and taking in too much water. This only occurs during extended and strenuous exercise, however, and is not a common problem during daily events.

Begin carrying your water bottle with you everywhere and keep the water coming in. You'll see and feel the benefits very quickly. Staying more hydrated will help your voice, your energy level, your digestion, your skin, and almost every one of your internal organs. All of these improvements will help you keep going during a crisis.

CAFFEINE

With a coffeepot in every newsroom and a Starbucks on every corner, coffee is so much a part of life that we rarely think about it. It is always available. A moderate consumption of caffeine (under 200 mg) may actually be helpful to our health and alertness. But too much caffeine from coffee, tea, or soda can be your worst enemy as a broadcaster, and it's worse when you are covering a crisis situation. Caffeine can make it more difficult for you to concentrate because it increases the adrenaline in your bloodstream. It also does not provide the needed moisture to protect your vocal tract, which may result in vocal damage. These problems are added to an additional long list of health problems that caffeine contributes to, including anxiety, insomnia, fatigue, indigestion, palpitations, blood sugar irregularities, irritability, and anxiety.

(By permission of Dave Coverly and Creators Syndicate Inc.)

So why do so many of us have to have our coffee? The answer is simple. Caffeine is an addictive drug. It starts a process in our bodies that perpetuates itself. I often describe caffeine as *liquid stress.* Our bodies do not know the difference between a cup of coffee and a man with a gun. At the first sip our bodies begin to pump out adrenaline. The book *Caffeine Blues* explains that "a single 250 mg dose of caffeine has been shown to increase levels of . . . adrenaline by more than 200%." This wakes us up for sure, but it does not give us energy; it gives us stress.

Caffeine just borrows energy from the future. Whenever we get a dose of adrenaline, we feel fatigue afterwards. So a short time after that java jolt, we feel worse than we did before we drank it. What cure is there? Have some more coffee or soda or tea. This creates a cycle of caffeine-induced energy followed by fatigue followed by the desire for more caffeine. We also do not get the depth of sleep that we need when we consume too much caffeine, which drives us right to the coffeemaker in the morning to provide the energy that a night of good sleep should be providing (see chapter 5).

The American Medical Association recommends no more than 200 mg of caffeine per day. The bad news is that a 5 oz. cup of coffee (and I challenge you to find a cup that small except in a baby store) contains around 115 mg of caffeine. Most caffeine we drink is poured into a 14 oz. mug or served as a 16 oz. grande or even a 32 oz. Big Gulp. That means what we call a cup of coffee is actually between three and six cups. And our caffeine intake may be anywhere from 350 to 700 mg after what we think is just one cup. Caffeine intake at this level affects our health and can have a strong negative impact on on-air delivery. Teas and sodas have lower levels of caffeine, but they are often consumed in greater quantity. Caffeine often hides in medications and places that you might not look for it. See Crisis Survival Kit 5 for a comprehensive list of caffeine sources.

Because caffeine is an addictive drug, never quit cold turkey, especially if you are in the middle of a crisis situation. Ramp down slowly as you limit your intake. Begin cutting your consumption in half every week by substituting decaf or halving your intake. Do this for as many weeks as it takes to get down to 200 to 300 mg a day.

Caffeine comes with a cost that is not worth it for most broadcasters. You are already working in a stress-filled profession, and you do not need to increase that stress chemically. This is especially true when a crisis breaks and your adrenaline is on overdrive anyway. Heavy coffee drinking brings on symptoms like jitteriness, shortness of breath, rapid heartbeat (possibly even abnormal heart rhythms), and upset stomach. These are the same symptoms that the stress of the business can bring on all by itself.

ALCOHOL, DRUGS, AND SMOKING

At times when stress becomes intense, such as during a crisis, we tend to reach for something to relieve that stress. Unfortunately, many people turn to alcohol, drugs, and cigarettes. All of these are poor choices for relief. It's a good idea to monitor your use of these items when you're dealing with the emotional and physical stress of a crisis.

If you drink to relax, be especially careful how much you rely on alcohol during a crisis. One network television producer told me that at first she really needed a glass of wine to unwind when she came home. Later that became two glasses, and finally she found she was drinking most of a bottle every night, and feeling the effects the next day. This is when she realized she had a problem.

According to the National Council on Alcoholism and Drug Abuse, it takes two hours for one ounce of alcohol to leave your bloodstream. Drinking too much one night will mean you are not functioning well for most of the next day. Your body will suffer, as will your appearance. As I heard one network anchor say, all the makeup in the world can't cover two margaritas the night before.

If you begin to depend on "happy hour" or prescription or illegal drugs, think about other ways to relax. Relaxation techniques and exercise are much healthier ways to relieve stress after a tough day. You may also need to begin talking about

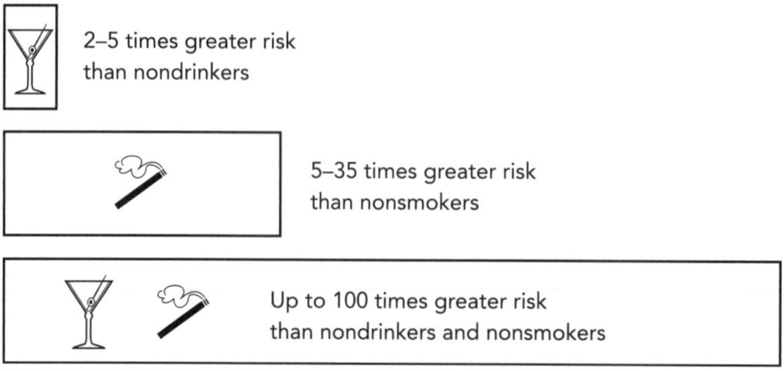

Figure 7-2 Relative risks of developing throat cancer (laryngeal and hypopharyngeal) for smokers and drinkers versus nonsmokers and nondrinkers.

(Source: American Cancer Society, Cancer Facts & Figures, 1999.)

your feelings. Monitor yourself and listen to your friends or family if they tell you that you are increasing your use of alcohol or drugs.

If you smoke, your use of cigarettes may also increase when you're covering a crisis. In terms of vocal health, cigarettes are the most harmful thing you can consume. Cigarettes bathe the throat in cancer-causing toxins. The American Cancer Society reports that almost all those who develop cancer of the larynx use or have used tobacco. Cigarettes contain chemicals that directly irritate the vocal folds when inhaled and exhaled. And it's not just the throat that suffers. In fact, since 1964 the surgeon general has identified smoking as the single most important preventable cause of death in our society. Smoking kills approximately *half* of those who continue to smoke. Alcohol consumption is another important risk factor. People who smoke and drink are at highest risk of cancer. If you feel you must smoke and drink, at least don't do them at the same time. As you can see from figure 7-2, this is a lethal combination.

The Bottom Line of Eating Well

1. Carbs supply blood sugar, which equals energy. Eat mostly complex carbs.
2. Protein renews and rebuilds our bodies and keeps our immune systems working well. Eat mostly complete proteins.
3. To distinguish carbs from protein, remember that carbs grow out of the earth and protein walks on the earth or swims in the water.

(Continues)

4. Aim for eating two-thirds carbohydrates and one-third protein at every meal or snack.
5. Practice grazing, or eating five small meals a day.
6. Always eat something that includes carbs and protein for breakfast, even if you're not hungry.
7. Drink at least half your body weight (lbs.) in ounces of water daily.
8. Limit caffeine intake to 200 or 300 mg a day.
9. Avoid turning to alcohol, drugs, or cigarettes during a stressful period.

CHAPTER

8

F = Focus on Fun and Normalcy

"I don't think of all the misery, but of all the beauty that still remains."

Anne Frank

One of the most difficult aspects of dealing with a crisis situation is that by its very nature it drowns out our usual reality. It's hard to put a catastrophe in perspective when it's bigger than anything we've previously encountered. Crisis redefines our view of reality. Prior to September 11, for example, no human being had ever seen two buildings over a hundred stories tall crumble to the ground. Our minds and emotions need time to assimilate something that unusual and grotesque. But, even in the midst of this redefinition, we need to focus on fun and normalcy in our lives to remain physically and mentally healthy.

The Chinese actually define health as "balance." This idea applies to everything you've read in this book, but it especially applies to focusing on fun and normalcy. Without pleasant experiences in our daily lives, we cannot have balance. The stress of a crisis often turns our attention away from pleasure and joy and focuses it totally on tragedy.

After any crisis we crave a sense of stability and familiarity, to counter the horror that we've experienced. Lee Woodruff, the wife of network correspondent Bob Woodruff, explained it this way in a June 2003 article in *Health* magazine: "After Bob was in Afghanistan there was a dullness in the back of his eyes that took weeks and lots of hugs, story times, and multiplication drills with the kids to erase. I think that a person can see too much hate."

Doing the mundane things in life can restore a feeling of balance. And doing fun things can restore a sense of joy. Often after a crisis, however, we feel such fear, or anticipate so strongly that another tragedy will happen at any moment, that we can't relax enough to do normal activities. There are several ways to overcome this feeling.

LIVE IN THE PRESENT

First, throw out the "when-then" thinking that tells you that when things are 100 percent calm and safe you'll relax and have fun. Chances are you'll never reach that "then," because life rarely feels that way. Instead, start where you are at the moment. Find some of the small pleasures that can cut through the fear and chaos. This might be a taste of cool water or the sound of the voice of someone you love. It might be watching your children sleep as Sherry Stalley did after September 11 (see chapter 11).

Another approach is to use the phrase "I'm okay right now." This is something I learned years ago when I had a fear of flying. The instructor in a course I took advised us not to allow our thoughts to jump ahead or to go back while on an airplane flight. His advice was to stay in the present moment. Take a deep breath and repeat, "I'm okay right now" as you exhale. This can keep you present and prevent panic. It may also allow you to feel some small pleasures.

Reality checks also help during a time of crisis. In workshops I did after September 11 and the D.C. sniper incident, I asked the groups to give me specifics about how their lives outside of work had changed in negative ways since the events. Most reporters and anchors realized that their lives had not really changed at all. They were still going to work every day, buying groceries, visiting with friends. Everyone felt that these events had changed their lives inalterably, but in truth their lives were much the same. Fear and anger were the only new additions, but nothing tangible had really changed.

Take a look at these statistics:

Lifetime Risk of Being Killed by . . .

Terrorism	Too small to calculate
Amusement park ride	1 in 920,000
Hurricane	1 in 220,000
Airplane accident	1 in 40,000
An intimate partner	1 in 1,800
Flu	1 in 1,700
Homicide	1 in 240
Car accident	1 in 88
Cancer	1 in 7
Heart disease	1 in 4
Tobacco-related illness	1 in 2 smokers

Primary Source: David Ropeik and George Gray, *Risk: A Practical Guide for Deciding What's Really Safe and What's Really Dangerous in the World Around You* (New York: Houghton Mifflin, 2002).

These statistics helped me not only with my fear of flying, but also after the attacks on Washington. It wasn't easy to think rationally when I saw workers in bioterrorism suits cleaning anthrax out of the building next to where I was working on Capitol Hill, but I kept reminding myself that my life was still normal. I had a much greater chance of being hit by a car crossing the street than being killed by inhaling anthrax.

PUTTING FUN BACK IN YOUR LIFE

If you read chapter 2 of this book, you already know how stress affects your body in negative ways. But here's one more example. Robert Sapolsky (see Suggested Readings) described a study in which actors were given a role to play for a day. One group was given a depressing role and the other an uplifting role. Just playing these roles actually changed the actors' immune responses; the actors who played the depressing role had a decreased immune response, whereas the actors who played the uplifting role had an increased response. If acting happy can improve our immune systems, imagine what feeling happy can do.

Let's look at ways to incorporate fun into your life when crisis strikes. First, start small. Simply smiling can stimulate the release of feel-good endorphins in our bodies. Now I'm not suggesting you put on some Pollyanna, pasted-on grin, but I do suggest you look for times when you can genuinely smile. This might be simply thinking of a loved one or a pet or smiling at a passerby, or it could be gazing up at the sky or looking at a tree. In the darkest hour of a tragedy, think of a smile as your lifeline.

Beginning with something as simple as a smile, you can start to parcel out fun and joy throughout your day. The small ways to find this fun are as varied as the people who pursue them. We all know what might feel good to us. Maybe it's chewing a big piece of bubblegum or looking at a flower. It might be petting a dog or cat or calling a friend. Watching babies and children is a great way to smile. Try making eye contact with infants you see and let yourself feel the joy of being alive. Thich Nhat Hanh, the Vietnamese Buddhist monk, calls these acts a way to diffuse the "bomb in our hearts." These small joys lift us out of the place of pain, even if just for an instant.

How can you tell when you need to do this? Here is a list adapted from the book *20-Minute Retreats* (see Suggested Readings). Answering "yes" to a few of these will show you that you need more joy and fun in your life.

- You can't remember laughing lately.
- You feel serious all the time.
- You're just trying to survive day to day.
- There are no celebrations in your life.
- You don't take vacations.
- You don't notice beauty around you.

- You don't like your job.
- You have no time for the important people in your life (partner, friends, children, etc.).
- You are turning to things that numb you (overwork, alcohol, drugs, excessive TV watching, etc.).

These are all warning signs that you're shutting yourself off to the good feelings in life.

The Healing Power of Laughter

Laughter often accompanies fun times, and that's one of the reasons fun is so important in life. Many studies show that laughter actually lowers stress hormones and gives us a dose of those feel-good endorphins. Bernie Siegel, a physician who is an expert on wellness, explains this very succinctly when he says, "When we laugh every cell laughs." Many researchers have shown that laughter can raise the immune system enough to have a profound effect on healing. So, let yourself laugh on a regular basis.

PUTTING NORMALCY BACK IN YOUR LIFE

We all perform activities in our lives that create structure and give our lives normalcy. They may seem mundane, but they actually make us feel safe and secure. These things include our mealtimes, the hour at which we go to sleep every night, and the people with whom we interact every day. During a time of crisis, it's important to keep doing as many of these things on a regular schedule as possible.

Crisis makes us feel out of control, and normalcy in our lives makes us feel in control. There are times when you're in the midst of a crisis that you may benefit from creating more control rather than relaxing. For example, if you're suffering from intense physical or mental pain, you can use a pain management technique in which you evaluate your level of pain every few seconds by assigning it a number. This gives you a way to feel in control, by comparing the numbers from minute to minute. In cases of extreme fear and anxiety you can use this technique to calm yourself. Or try looking around and observing what is still normal. Look at the sky. Watch the birds. Observe others around you.

When you are not actively covering the crisis, you can restore a sense of normalcy simply by doing normal things again. Go to the grocery story. Make your bed. Call a friend. Play with your kids. Do any activity that's a part of your everyday life, and really take in that things have not changed as much as it feels like they have. Remind yourself that life will not always feel as intense as it does during the crisis.

TAKE ACTION

Very often in the news business you spend your time reacting instead of taking action—especially during a crisis. The crisis happens and you and your coworkers respond to it, and you can easily get caught up in simply responding to what's happening around you. However, in order to have fun and create normalcy in your life, reacting isn't enough. You will have to *act.* You will have to set aside some time to have fun and engage in everyday activities.

If you grew up like I did, you may think that your job on the planet is to take care of everyone else first, and if there's a scrap of time left over maybe you can do something for yourself. I can tell you that such thinking is backwards. It's not selfish to take time for yourself. In fact, if you don't take this time, you won't have anything to give others. Again, think of the analogy of the airplane emergency rules: put your oxygen mask on first before trying to help someone else with his. If you're not being replenished, you can't help anyone. Broadcasting through a crisis requires lots of giving on your part. Allow yourself time to be recharged. Alice Domar, director of the Mind/Body Center for Women's Health in Boston, suggests we think of this concept as supply and demand. For example, if your job demands that you cover a crisis for ten to twelve hours in one day, you certainly deserve to spend your off hours focusing on you.

If you tend to keep a scanner or the news on when you're away from work, think of limiting your exposure during crisis coverage. I know this is hard, because you don't want to miss important developments, but plan to escape for an hour or two at least. Pain researchers at Stanford University found that just watching someone being hurt or suffering through a crisis triggers brain activity that is similar to actually being hurt. If you constantly expose yourself to coverage of the crisis, your body will react as if you were still in the midst of it yourself. You need some down time, no matter how big the event.

SCHEDULE FUN

To prepare for times of increased stress, I suggest you find something you love to do and begin to schedule it into your life now. Consider it as necessary as getting to work every day or eating at the right times. Don't let anything get in the way of your fun time once it's scheduled. If you begin this process when you're not in a crisis, chances are you will keep it up when the crisis hits.

Many people on their deathbeds say that extra hours they spent at work now mean nothing to them. What they remember is their son's homerun that they saw because they left work to go to the game. They treasure the times with family and friends. When you read the Profiles of Crisis in this book, you'll see that almost every reporter talked about how much family and friends meant to him or her. These are the meaningful moments that provide normalcy in our lives. To quote Sophocles, "One word frees us of all the weight and pain of life: that word is love." Schedule time with loved ones and really take in their love.

When a crisis hits, it may be a time to *pay* for some help with your life so that you do have time to have fun. You might consider hiring a housecleaning service or a babysitter to give you time to replenish yourself. Take some of the burdens out of your life for a few weeks. Even when the crisis ends, it's really helpful to put yourself first at least for a while so that you can stay physically and emotionally stable.

Ways to Have Fun

You are the best expert on what you like to do to have fun. Here's a list of some of the things clients have told me they turn to when they are stressed and need to schedule some fun:

- Listening to music
- Watching a funny movie or TV show
- Dancing
- Playing sports or watching sports
- Meeting friends or family for coffee or a meal
- Playing with children
- Observing the beauty of nature
- Gardening
- Pursuing hobbies such as painting, playing cards, or working on model trains or planes
- Hiking
- Boating or being around water
- Attending a theater performance, ballet, opera, or concert event
- Visiting a museum
- Riding a bike
- Playing with animals or going to the zoo
- Going to a toy store and watching kids play
- Spending time in a hammock or simply sitting and watching the clouds or stars

You could expand this list to include anything from bungee jumping to knitting, and there are hundreds more ways to schedule some fun in your life. Now is a good time to write down some of the ways you like to relax and have fun. Keep your list handy, and incorporate those activities in your life as often as possible.

Another way to feel better when covering a crisis situation is to find ways to soothe yourself. Self-soothing calms us and makes us feel good. Activities like taking a warm bath or shower, massaging muscles, stretching out on the floor or a bed or couch and relaxing, or listening to calming music make us feel good and lower our anxiety. Anything that does not include alcohol or drugs or any overindulgence (like a whole pint of Ben and Jerry's) is fair game. Self-soothing provides a way to release some adrenaline and relax, even if just for a few minutes. Add some self-soothing activities to your list of ways to have fun. You'll find other examples of self-soothing activities in chapter 3.

GOOD THAT COMES FROM TRAGEDY

As the quotation from Anne Frank at the beginning of this chapter illustrates, no matter how awful life seems, beauty still remains. Focusing on what's good in your life can show you that life is more than crisis, anger, or sadness. While researching this chapter, I came across the phrase *post-traumatic growth* several times. This is a concept that really sums up the good that can come from tragedy.

I think we all noticed post-traumatic growth after September 11. We saw how communities came together to support each other. I certainly witnessed this first-hand when my husband and I had to drive across country to return home since the D.C. airports were closed. Here is a section of a piece I wrote that aired on national radio:

> I wish each of you could have seen what we saw. First, in every little town and on many overpasses we saw flags draped everywhere—on houses, on bridges (many with handwritten posters beside them), on highway equipment, and even on a very disheveled hitchhiker who had a flag draped across his back. On some overpasses, groups of people were standing and waving flags at the cars passing below. And at every truck stop, the electronic signs that usually show gas prices said, God Bless America, or United We Stand, or In God We Trust, or May God Be With Us. Many cars had flags in their windows or hastily made computer signs with flags. Many trucks had full-sized flags attached to the doors. The patriotic response we saw for over 1000 miles was overwhelming. We are a stronger nation and stronger people having come through this.

Shortly after September 11, *Newsweek* reported that 80 percent of Americans believed that the attacks would make our country stronger. Even if the positive response didn't continue with the same intensity we felt right after September 11, our country and all of us were changed for the better having experienced the positive aftermath of the tragedy.

The same kind of post-traumatic growth was seen after a less tragic event, the power blackouts of summer 2003. The crime rate actually went *down* during the night of the blackout in New York City, and people came together in a new way as they camped out on the sidewalks. All across the East Coast and the upper Midwest people helped each other get through the outages. This same spirit was seen when four hurricanes hit Florida within weeks of each other in 2004 and following the horrific tsunami at Christmas of that year.

What these positive experiences show us is that tragedy can bring out good in people. Good can come from tragedy because we always have the choice of how we view an experience. When we choose to see an event positively, we experience post-traumatic growth.

A powerful example of this is described by Viktor Frankl in his book *Man's Search for Meaning* (see Suggested Readings), when he relates his experiences in

Nazi concentration camps. He explains that the prisoners who gave meaning to their experiences were more likely to survive. He concludes that, "Everything can be taken away from a man but one thing: the last of human freedoms—to choose one's attitude in any given set of circumstances, to choose one's own way."

What Frankl witnessed was that the horrors were the same for all Nazi prisoners, but the way they thought about their day-to-day experiences affected their survival. It appears that, even in hell, there are choices. As John Milton said in *Paradise Lost,* "The mind is its own place, and in itself can create a Heaven of Hell, a Hell of Heaven." The choice is ours.

Allow yourself to remain optimistic even in the face of tragedy. *The Post-Traumatic Stress Disorder Sourcebook* (see Suggested Readings) defines optimism this way: "It is the attitude that no matter what happens, I can find something to enjoy. It is the choice to be happy despite obstacles." Aristotle said this same thing over two thousand years ago when he wrote, "Happiness depends on ourselves."

Every action we take is colored by how we view it. We can see the dark side or the light. We can feel gratitude or pain. We can experience joy or stay in the horror of the crisis. Focusing on fun and normalcy in your life will allow you to see the good in any situation. Then you can reach the place that Albert Camus described when he wrote, "In the midst of winter, I finally learned that there was in me an invincible summer."

We always have a choice in how we feel. I am reminded of this when I hear this song:

> When I seek, the world is my teacher.
>
> When I worry, the world is my burden.
>
> When I give, the world is my garden.
>
> When I hoard, the world is my desert.
>
> When I reach out, the world is my family.
>
> When I strike out, the world is my enemy.
>
> When I trust, the world is my harbor.
>
> When I fear, the world is my storm.
>
> When I love, the world is my heaven.
>
> When I hate, the world is my hell.
>
> What will my world be today?

Written by Fred Bogert
From the CD *Dreamspeak*
by Fred Bogert, available at
www.chromerecords.com

The Bottom Line of Focusing on Fun and Normalcy

1. Focus on the present and remind yourself that you're okay right now.
2. Recognize the real risks and how likely they are to happen.
3. Create balance in your life by scheduling time for fun.
4. Find ways to laugh!
5. Create balance by continuing to do normal things every day.
6. Parcel out small pleasant experiences throughout your day.
7. Recognize that it's not selfish to put your self-care first during a crisis.
8. Limit the time you spend time watching the news, etc. when you are away from work.
9. Practice some self-soothing activities.
10. Look for the good that may come out of any bad experience.

PROFILES OF CRISIS: REPORTERS IN THE TRENCHES

"They say time heals all wounds. I'm not convinced that's true. . . . I'm not sure it gets easier; it just gets less hard."

Mike Walter

"I honestly think we put aside being 'reporters,' in the true sense of the word, and instead we were humans who had to experience and tell the world about stuff no one should have to hear."

Sherry Stalley

CHAPTER

9

September 11 at the Pentagon

by Mike Walter, WUSA-TV, Washington, D.C.

September 11, 2001 was a day rich with irony. It began as a bright sunshiny day, yet it would end up being the darkest day of my life. It was also supposed to be an easy day. I had celebrated my birthday on September 1. My son had celebrated his on August 31. We were both looking forward to September 11 because we would share a belated birthday gift together. We were going to an Orioles baseball game in Baltimore. I planned to get out of work early and return home to pick him up. When I left the house that morning I had no idea I'd never make it to the game, and by the time I got home my son would be asleep.

A coworker once said I always looked like I was on a job interview, because I am always in a suit and tie. But this was one of those rare days when I would leave the suit and tie at home. I gladly put on a polo shirt and khakis. As I left for work I couldn't help but think today would be a breeze. No television live shots for the twenty-two Gannett stations today. I would get out of the office early. I was wrong on all three counts. It would be one of the most difficult days of my life. By the end of the day I would be on the Gannett stations as well as all the major networks. The hours would be long, and stressful. Most of the hours I worked in a state of shock, at times in a state of terror, always in a state of disbelief. It would be days before I would feel like I was emerging from a deep funk.

There was one other twist that day. Normally I make it into work on time or early. But on this day I wouldn't make it into the office at nine thirty. Because of that final twist of fate I would be in position to see the devastating effects of what has become my generation's Pearl Harbor.

When I think about that morning I think about the old man in the maroon car. Where was he when the plane struck the Pentagon? What did he see? Chances are he missed it all. I spent much of my morning trying to maneuver around him. He was slow and plodding. I was harried, and in a hurry. All I wanted was to get around him, but I was pinned in. The man in the maroon car was my nemesis. Finally I found an opening and was able to weave in and out of the lanes. Before long he was a distant speck in my rearview mirror. What if I had just stayed behind him? I might have slept a lot better that night, and the nights that followed.

But I was a broadcast journalist listening to NPR. There was a huge story taking place, and I was trapped in traffic. It started as such sketchy reporting. A plane had hit the World Trade Center, and then later a second plane had struck. Now both of the Twin Towers were burning. What was going on? All I could think was that the radio was the worst possible medium for me. What I needed was a television. If I were at work I could be watching the events unfold and figure out things for myself. The radio announcer grappled to find words to describe what was going on, without much success. Suddenly the President was on and calling this terrorism. I was growing more and more anxious. I wanted to get into work but the traffic still wasn't moving.

I slammed my hand on the steering wheel, grunted, then finally took a deep breath and rolled down my window. Just as I did this, I heard the roar. It was the sound of a jet engine. I looked up to see the sleek silver fuselage. The jet banked over my car. Now instead of thinking about traffic and what was happening in New York, I was fixated on the plane. It was low, very, very low. It was as if the pilot was lining up the Pentagon. Once pointed in that direction it sped up, dipping lower. I saw the big "AA" painted on the jet. It dove. I watched in disbelief as it began screaming toward the side of the concrete structure. It was a strange, helpless feeling. I knew the plane was going to crash. I know it sounds clichéd, but it did seem like time was suspended, that all of this was unfolding in slow motion.

I decided I needed to keep my eyes glued to what I was watching. I was a reporter, and that's what I'm paid to do, be observant. I knew the outcome before it happened. I still had to wait for this tragedy to unfold. The jet dipped out of sight for a time, after passing over the highway I was on, then reemerged and pierced the side of the Pentagon. The jet folded up like an accordion as it continued to plow into the building. There was an immense fireball, followed by a huge column of black smoke. I was in a state of shock. Chills ran up my spine. I mumbled over and over again, "Oh, my God. Oh, my God." Then I said, "I can't believe this. I can't believe this." There was no one in the car with me, but I still said these things out loud.

Other drivers panicked. A woman in front of me in an SUV shouted, "Get back, get back. They've hit the Pentagon. They've hit the Pentagon." Like me, she was saying things twice. Because no one was listening to her she drove her car forward, then threw it into reverse. She was frantically trying to free herself from this tight squeeze. Finally she managed to turn her car around and she drove the wrong way back up the highway, desperate to get away. At the same time other cars were racing either forward or backward in the emergency lanes. I pulled my car over and got out. I started to make my way to the Pentagon.

I had been only seventy-five yards away from the point of impact when I saw the jet slamming into the building. Now I was getting closer. I looked down and saw pieces of the wreckage sprinkled on the asphalt. I thought to myself, if this is all that's left of a jet, what happened to the people on board? I already knew the answer. I gingerly stepped over the wreckage. I was awash in sights and sounds. The soundtrack to this day was a siren, on an endless loop, sprinkled in with the sound of sobs of the people who had escaped the inferno. The sights, the unbelievable fury of the fire as it continued to chew its way through the building. Other sights: An endless stream of fire engines and ambulances racing to the scene. Military personnel setting up a triage area. People in uniform racing to the scarred side of the Pentagon hoping to retrieve friends and colleagues.

Later still the sight of terror in the faces of officers and soldiers as they screamed at us to leave the area. Another jet was headed our way and it would be here in fifteen minutes, they said. I remember the terror in the faces of the bystanders loitering about, as they raced to get away. Jets were raining down all over the East Coast and another deadly storm was headed our way.

All of these sights and sounds added up to sensory overload. But I had a sense of purpose that day. I wanted to find a photographer from our Gannett flagship station, WUSA. If I could go live and tell viewers what I had seen then there would be a reason why I was there witnessing this horror. But each videographer that I stopped was from another outlet. Finally I found a photographer. I introduced myself. "I'm Mike Walter with *USA Today Live*." He showed little interest. He had a job to do and I was some guy bothering him. He moved away. I took a deep breath and decided if I had to be a nuisance I would be.

"Look, I'm not trying to bother you, but I witnessed the plane going into the building. I just want to give you guys my eyewitness account." Suddenly his tone changed. He directed me to anchor Gordon Peterson, who was some distance away. It meant that I would have to scale a fence and scramble down a hill to get to him. I started up the fence and got to the top when I got caught. The metal tore a gash in my right wrist about an inch long. I lost my balance somewhat and managed to get to the ground on the other side. I was bleeding and in pain. I must have looked pretty pathetic as I staggered to get to the live shot location. Just about out of breath, and operating on adrenaline, I reached Gordon.

After the interview I had to wait for my own photographer, colleague Lance Ing, to arrive. I pride myself as a good writer for television. That means looking at images and writing to them. Here were all of these incredible images unfolding in front of me. But there was no one there to shoot them. It was one of the most frustrating experiences I have ever encountered. Hours later, after bouncing from roadblock to roadblock, Lance arrived. By this time I had agreed to do a number of interviews with network reporters. John McWethy of ABC was the first to interview me. Jamie McIntyre of CNN was the last.

Throughout the day I was borrowing cell phones to try to contact my family. Most of the cell phones were useless. People stood in long lines to use a gas station's only pay phone. On this day new technology would take a back seat to old. Loss was a big part of that day; there was the loss of life and the loss of control, and in many ways the worst part was the loss of communication.

By the time I walked over to talk to Jamie in the afternoon, I had finally made contact with my family. Before going live on CNN I had talked to my teenage daughter. She was a wreck. So many of her friends had parents who worked at the Pentagon. So many of them were still unaccounted for. What do you tell a teenager when you can't make sense of what's happening? It's the role of a father to provide some sort of wisdom or comfort during these encounters. But I had nothing I could say to her that day other than there's a good chance those kids will never see their parents again. But of course I couldn't say that.

I certainly thought it as I watched the flames move from room to room of the Pentagon, devouring anything in their way. Were the parents of some of my daughters' friends in the way of those flames? I would learn only later that the answer was no, and all my daughter's friends had checked in to tell her their parents were okay.

I fought back tears as I did the live shot with CNN, and it was one of the hardest parts of that day. The last interview of the day was conducted by my friend and colleague Lance Ing. He interviewed me, and I recounted again what I had seen. When it was over, and we were getting ready to go back to the office, I told him, "I'm happy this day is coming to an end. I'm happy this is over." He laughed and said something like, "You know better than that. You know this business; you'll never be over this. There's always a six-month anniversary, and a year anniversary and a five-year anniversary. People will be interviewing you about this for years."

When I finally got home that night, I had trouble sleeping. I managed to get a few hours of sleep before heading to downtown Washington for live interviews on CBS and CBC. Traveling on the freeway I looked out the window and through the darkness spotted the orange ball of fire still chewing through the Pentagon. A thick blanket of smoke enveloped the freeway. Washington was a ghost town. There were virtually no cars on the road that morning. By the time I hit downtown it was as if I had entered another country. Soldiers clutching weapons stood on almost every street corner. It was spooky. I did the live interviews, then went to my office where two FBI agents greeted me. They wanted to know what I had seen. I spoke to them for about fifteen minutes.

Growing up in Southern California, I had lived through earthquakes. In many ways this was the same. When the quake hits it comes as a surprise. But most unsettling are the days that follow, the uncertainty. There are the aftershocks and they can come at any time. You are on edge. When will the other shoe drop? I had those same feelings in the days following the attacks. We waited for the terrorist aftershocks. What's going to happen next? Are we safe? We didn't know the answers to either of these questions. Our minds raced and conjured up every horrid scenario imaginable.

Work and home life blended together. Both were consumed with watching the images on television, or creating images for television. When I wasn't out working on the story, I was watching it unfold on the air. It all managed to sink me into a deep depression. In the days following the attack, tears were like balls in a pinball machine. I would start to cry, get a grip, and then look over at a producer who would be in tears; by the time she was done the tears had moved on to the next person in the office. Each of us had our own reasons for crying. One of my coworkers had family and friends in the New York area. Another had a husband who had ties to some of the financial planners and brokers who died in New York.

My boss suggested that I see a counselor. I didn't want to take time from work, so I decided to do a story on the group counselor who was working with employees at my place of work. I went to the session and shared what happened to me. I thought this would be seamless and easy. I would share what I had seen. Naturally the counselor would wave her magic wand and I would be better. That never happened. What happened instead? Well, I told my story, and then listened to others share their stories, each more gut-wrenching then the last. I just wanted to unload, but I left with an even deeper sense of dread. I hadn't lost my pain, and it seemed like I had absorbed everyone else's. I went to another group session with the same results.

The emotional fallout was also being felt at home. I was obsessive. I craved coverage about September 11. I would devour television coverage, and in the morning I was up early to read the newspaper accounts. I remember one morning sharing one of the stories with my wife. She suddenly said, "Oh look there's a big sale going on at Kmart." I was very upset. "How can you talk about that, how?"

A couple of days later she suggested we get out of the house and go for dinner. We did. It was at dinner that she said, "Don't you think this is eating me up too? But someone in the house has to hold it together. All this talk about what happened is scaring the kids. It's all you think about. It's all you talk about. They need to know that life is normal again, or at some point it's going to return to normal."

My wife was the glue holding things together in our house. But life wasn't normal. I felt distant. I became consumed with what happened. "Why?" is the question that tortures you. Why me? Why did I have to see it? Why did it hit there—what if it had crashed before the Pentagon? If it had I wouldn't be here. Why did I survive, while so many others died? The time I dreaded most was when I was alone. Even when I was alone in the house, or alone in my car, I wasn't really alone at all. The images would always be with me. They were always there at night. When I was able to fall asleep the nightmares would be there. The nightmares gradually faded over time. But in the initial days after the attack, the obsession with coverage, the images that haunt me day and night become my bricks and mortar. They form a wall that keeps me detached and distant. The terrorists have robbed me of who I once was. I'm disconnected. I feel like I'm in my own little world, unable to connect with anyone who didn't experience what I did that day.

On the one-year anniversary of September 11, the nightmares returned. The restless nights came back, filled with tossing and turning. The stories were creeping back into the papers and back on television. The news coverage had been like an I.V. in the hours and days following the attacks. Somehow I was transfixed. It was as if I had to watch. I had no choice. Now it was different. There was a sense of dread. I agreed to do some interviews. I had my reasons for this. We must never forget what happened that day. For those of us who were there, we have an obligation to keep the story alive.

So what's it like living through that morning? What's it like living with those haunting images? They hide in the recesses of my mind, popping up from time to time like an uninvited guest who refuses to leave. The American Airlines jet dropped from the sky and raced to its destination of death. But it will never drop from my consciousness.

I attended a conference in Miami where they talked about reporters dealing with Post-Traumatic Stress Disorder. The conversation focused on September 11. All of the symptoms described fit me to a T.

I feel like I've lived two lives now. The one prior to September 11, and the one after. I know that I'm a different person now. But I've taken steps to move on. I've gained a great deal of strength from my family. When you see lives extinguished in a matter of seconds, it helps you realize how precious and how fleeting life can be. Priorities change. After the attack I decided to put a premium on my time with my family. I also put a premium on exercise. I attended church before, but my faith has been strengthened by what happened.

They say time heals all wounds. I'm not convinced that's true. There are signs that I'm healing from the wounds of September 11. The scar on my right hand is fading. The emotional scars of that day are as well. But they will never fade entirely. I'm not sure it gets easier; it just gets less hard.

The further I get away from that day the more I try to look at the good. In many ways I think there are good things that came out of that day. One of the guys I interviewed on the second anniversary of September 11 who was also involved in the recovery effort said it best: "You have to move on. If you let the past consume you, it will destroy you." As he put it, "Learn the lessons that need to be learned and move on." I think I have, or am.

Mike Walter is currently the morning anchor at WUSA-TV in Washington, D.C. Prior to joining WUSA, Mike was senior correspondent at *USA Today Live.* After September 11 Mike appeared on *The Oprah Winfrey Show,* and he told his story to journalists from Britain, France, and Germany. Mike has been an anchor at local television stations in Tampa, Kansas City, and Columbus, Ohio, and two stations in California. His reporting career has taken him to the war zone in Somalia and to the Kremlin in Russia, and he's also covered earthquakes in California, tornadoes in Kansas, and hurricanes in Florida.

CHAPTER

10

September 11 in New York City

by Melinda Murphy, CBS News, New York City

It's been many months since I covered this event, but the mere mention of it still brings back a dull, sad ache.

Unlike some of my fellow reporters and producers, I don't avoid it. I know plenty of people who refuse to watch any more video or read any more articles or even write about it. But, for me, writing is a therapy, of sorts.

I was the last airborne reporter aloft over New York that day. At the time, I was working for WPIX-TV in New York. I was the traffic and feature reporter for the morning show and was kind of the nut on the show. That's not to say that I didn't do breaking news. I did. In fact, I was nominated for an Emmy for my coverage of a fire just that year, but most folks thought of me as the show's designated goofball.

On September 11, I put back on my "hard-news reporter" face.

We had already finished our morning show, and I was back at the station along with my photographer, Chet Wilson. When I saw the first images on one of the television monitors in the newsroom, I screamed at Chet and the two of us went flying out of the newsroom. As we left, we called our pilot and asked him to launch the helicopter.

We tackled a cab and made our way from Forty-second and Second to the heliport on the West Side of Manhattan. My husband worked a block south of the towers and he was the one who told me that a plane had hit the World Trade Center when I called him from the cab. I told him I'd be home late. I had no idea what we were about to all face. How could I?

Our helicopter was just landing as we pulled up to the heliport. Ray Rivera, our pilot, had made his way from Caldwell Airport and saw the second plane plow into the tower. As we took off, I remember just being awed at the sight of this huge fireball in the air. I had covered a lot of fires from the air, but this was like something I had never seen. I couldn't find the words to describe it even to myself, let alone to the viewers. I still can't.

We went live almost immediately. The impact had knocked out the station's signal, and I couldn't see or hear my own station, so I wasn't sure what the anchors were saying and if I was in line with them or not. On top of that, nobody could ask me any questions. I just had to talk. My first on-air report was eight minutes. That's a long time to talk and make sense out of something that made no sense whatsoever. And, when I listened to it later, I realized I somehow did pull it together.

Luckily, I was in the air with two of the best people I've ever known. Chet had been a firefighter and kept telling me all this really helpful stuff about fires. He had already taught me a lot, but this was different and he really was a great resource. Ray was also feeding me information about what was going on with air traffic. We were a team in the truest sense of the word. Each of us was an equally important link to getting the story on the air.

After our first initial reports, things slowed down, and I was finally able to start to really think about what was happening. Up to that time, I had been on autopilot. I never stopped to think about running toward the scene. I never stopped to think about jumping in the cab. I never stopped to think about reporting what I saw. It was instinct. There was no question. It's what most journalists do. And it's not because we have some greater calling or because we are trained to do it. It's something I can't explain. But I knew I had to be there.

And, in those few minutes when we caught our breath, it started to sink in. Lots of people were going to die. Some were already jumping out of the windows, and we could see them from our vantage point. It was really, really shocking. I kept thinking about that reporter who covered the *Hindenburg.* "Oh, the humanity."

We had a faint signal from some of the other stations, and we started to hear about Washington and Pennsylvania. That's probably when I started to get scared. I was never terrified. It wasn't like that. I just kept wondering what was happening to our country. I wondered about my friends who worked in the buildings. I worried about my friends who would be up close covering it from the ground. I remember being consumed by guilt that I wasn't on the ground doing more. I felt so helpless in the air. The truth is, I was probably able to report more from the helicopter than I would have been able to do on the ground, but I still felt really guilty about not being in the thick of it.

Chet predicted the first tower would fall. I remember screaming at him, "It is not going to fall!" The thought was just too overwhelming.

And, then, it did.

There are no words to describe that moment. The whole thing happened right in front of our eyes. Floor after floor pancaked down. This huge smoke and dust cloud covered all of Lower Manhattan. For the first time in my life, I was completely speechless. The helicopter got very quiet. None of us talked. There was just the sound of the rotors. And then, this terror took over.

"My husband works right there." I said it out loud with a voice I didn't even recognize. I don't think I was saying it to the guys as much as I was saying it to myself. It just tumbled out of my mouth.

My husband Tom worked a block south of the buildings. When we talked last on my way to the heliport, he told me he was going to go outside to get a closer look. All I could think about was whether or not he was safe. Had I told him I loved him when we last spoke? And then I started to think about all my coworkers. I was shaken to the core.

I also remember wanting to talk to my mother. She'd died nine months before, and I just longed to have her with me. We'd been very close, and I really wanted to hear her voice, yet a big part of me was glad she didn't have to see it all. Still, the sadness of the day made me miss her all the more.

But I didn't cry. I teared up. I was stunned. I was overwhelmed. But I never cried. At least not yet. . . .

And then my associate director told me they wanted to come to me. Somehow, I snapped together and gave another long report. I just went into some kind of automatic mode and kept going. Working made it easier. I couldn't do anything from the air, so I might as well work. Now and then, I asked my producer if she'd been able to reach my husband. The answer was always "No."

When the second tower fell, it was more of the same.

We finally had to land for fuel. This was probably around one o'clock or so or maybe later; I'm not really sure. Everything from that day just kind of runs together. The other helicopter crews traded stories with us. I tried to call my husband, but I couldn't reach him. Cell phone service was pretty much nonexistent.

I called my elderly father, who had recently moved from Texas into assisted living in New Jersey. I was still holding it together until he said to me, "Where's Tom?" Then the dam just broke. I started sobbing, telling him I didn't know. I started telling him about all my friends and coworkers. I don't know if I even realized how worried I was until that very moment.

As we were about to climb back into the chopper, my cell phone rang. It was my sister-in-law, who worked across the street from my husband. She sounded strange. It was probably shock. She asked me to hold on. I thought she was trying to tell me my husband was dead, but then I heard his voice. My knees gave out, and I was on the ground just sobbing from relief. It was like something out of a bad movie. I remember my photographer giving me the thumbs-up or thumbs-down sign, trying to figure out if I was happy or sad. We talked very briefly, and I climbed back into the chopper to continue working.

We were finally asked to land for good at four forty-five that afternoon. For hours, we had been the only aircraft flying over the entire country that wasn't a rescue or military vehicle. We were using our strong camera lens to identify other aircraft and gather what news we could. At one point, air traffic control warned us that if we made any sudden movements they would shoot us down. It was pretty terrifying. I stupidly asked for my life vest (the one I wore when we flew over water). Like a life vest would help me if we were shot down!

And, through it all, I managed to keep it together. Or at least on the air I did.

When I finally got home that night, I hugged my husband for a very long time. But the truth is, he wanted to hug longer, but I just couldn't. I think I was so relieved to see him, and at the same time suddenly terrified to be close to him in case I'd lose him some time in the near future. It was a weird tornado of emotions.

That night, we had a rainstorm and there was thunder. I only slept for a couple of hours before heading back to the office (I was only home maybe three hours total), but when the thunder clapped in my sleep, I awoke with a start, certain we were under attack again. That's when I began to realize that I was pretty tightly wound.

Over the next few weeks, I worked impossible hours, but I was never sent to Ground Zero. I was the only traffic reporter for my station, and the attacks had changed traffic patterns drastically. Bomb threats were closing roads left and right. It was pretty intense. I slept only in snatches of time. I was so tired, sleep came easily, and I don't remember dreaming at all.

I found my greatest help was work and writing about what I'd been through. I was still consumed by guilt that I hadn't been on the ground like all of my fellow journalists. I kept thinking that's where I belonged. I probably had more air time than a lot of other reporters during those first few weeks, and looking back I think I probably imparted some pretty valuable information, but I didn't feel like I was really contributing. And that feeling never left me, even after I was nominated for an Emmy for my work that day.

I finally went to Ground Zero on my own about three weeks after it all happened. I was recognized by one of the police officers on scene, and he took me right down to the actual site. I was just overwhelmed with grief. Tears dripped down my cheeks. And again I was consumed by the feeling that I should have been there on the scene.

In the weeks following the events, the helicopters were banned from New York airspace. The first time we were allowed to fly over the city again was on Thanksgiving. I wasn't expecting to be routed over Ground Zero and was unprepared to do so. But they routed us on the exact same flight path as the last time I'd seen the Twin Towers intact. They'd been bathed in crimson sunlight on the early morning of September 11. They were the prettiest I'd ever seen them. Now the footprint looked like my worst nightmare. Metal was twisted and still smoldering. Fires were still burning. It was just unbelievable and much more horrifying than I'd imagined it would be. I looked at both my pilot and my photographer, and they were on the verge of tears. So was I.

My heart ached for those buildings, especially that morning. We'd had a special relationship with them. In the air, we used them to help establish where we were. They'd been our guiding light of sorts. And they were gone.

It took me a long time to get used to that. Even now, months later, I look at the New York skyline and miss those buildings. I wonder if I always will. I didn't lose anybody that day. I just lost two gleaming buildings. But I know the sadness must be over what the loss of them represents, not the actual structures themselves. And, of course, the loss of humanity.

I guess my therapy of sorts was working on a book about the events in 2002. I believe it is one of the best accounts of the day, *Covering Catastrophe: Broadcast Journalists Report September 11*. It's a play-by-play account of the day told through the eyes of more than one hundred journalists who covered the events in all three cities.

Putting that book together was incredibly tough. Every day for months, I read people's firsthand accounts. I was forced to talk about it, to think about it. But I'm proud of that book. And I'm proud that all our monies went to charity.

But there were things about that book that were hard, too. First off, reading the accounts of all those folks on the ground just heightened my own guilt for not being there myself. I know that's a ridiculous reaction, but it's what I felt—and perhaps still do.

What caught me most by surprise was the one-year anniversary. I thought it was all behind me. I thought I had found a way to neatly tuck this story away just like all the other hard stories I'd done in the past. The tucking perhaps took more time, but I'd done it.

But, on the one-year anniversary, I put together a piece about our day in the helicopter. Because of the book, I'd talked a lot about the events of the eleventh and my experiences that day. I'd also listened to lots of my taped reports. What I hadn't listened to were the conversations I'd had with my crew when I wasn't reporting. Those conversations were next to impossible to hear. And the hardest of all was when I'd said out loud, "My husband works right there." Hearing all those random conversations with my crew really opened up old wounds.

One wonderful thing that came out of that day is a very true and strong bond that I now share with both my pilot and photographer. We'd already been friends, but surviving those events together tied us together for life. No matter where we are, we will always be dear, dear friends.

As for my other relationships, I'm not sure how they were affected by the day. It's too hard to see how it's changed things for me, though I suspect that for a while I kept a greater distance between my husband and myself than I might have otherwise. We hadn't been married that long when it all happened, and I think I realized that day how much I really do love him—and how much it would hurt to lose him.

I do know one thing for certain: the events of September 11 will live with me always. And, as a journalist, I'm much more aware of what I might face in the future. But, for the most part, I don't think about it. I don't dwell on the attacks or that day. I just move ahead day by day with a greater appreciation for what I have. And I know more than ever that it may not be there tomorrow.

Melinda Murphy's stories about the lighter side of life can be seen on *The Early Show,* where she has been a correspondent since 2002. Prior to coming to CBS News, Murphy was a helicopter/feature reporter with WB11 in New York City and a live feature reporter at News 12 New Jersey. She has won several Emmys, but is most proud of receiving a nomination for her coverage of September 11.

CHAPTER

11

September 11 in Pennsylvania

by Sherry Stalley, FOX 8 News, Johnstown, Pennsylvania

The morning of September 11 started as any other day in the newsroom. The only exception was that it was my first day back at work after a five-month maternity leave. Ironically the day started very slowly. As we all know, a slow day means extra work trying to find something to fill the newscasts.

Shortly after eight that morning someone blurted out, "Way to come back to work, Sherry, when nothing is going on." Oh, how wrong they were. In less than a minute our eyes became fixed on the television monitor. We saw breaking national news. A plane had apparently just crashed into the World Trade Center. The morning producers were happy because they finally had a lead. It wasn't until we all watched as the second plane crashed into the other tower that we realized we had local angles to cover. There was talk of possible terrorism.

Localizing a national story is very important for smaller communities like the Johnstown area. Immediately I was on the phone to our 911 center. I knew they had just enacted local protocols in case of a disaster, and I thought maybe this would fall into that category. I was right. Security was immediately stepped up at several places in our community, including our federal building, which houses the National Drug Intelligence Center. I set up an interview with the emergency management director, and my photographer, Keith Hoffer, and I jumped into the live truck and were on our way to do a live shot for the noon news. From there when I look back it all really is a blur. So much happened that I think part of me doesn't want to remember. As a reporter, however, I can still remember every detail.

Our 911 center is about twenty minutes from the station. We were just a few minutes out when our scanners went crazy. Talk of a plane headed to Johnstown with a bomb on it. Talk of another plane down in Somerset County. Desperate pleas for help from every available police, fire, and EMS worker. There were emotions coming from dispatchers that in my twelve years on this job I had never heard.

It was at that moment that Keith and I literally stopped in our tracks in the middle of the interstate. We just looked at each other, not knowing where to go. Our cell phone wouldn't work. More information began pouring out of the scanner.

There was talk of evacuating the city! And there was more on that downed plane. Prioritizing stories took on a new meaning right then and there. What was more important? What impacted more people? I grabbed the phone and prayed it would work. It did. I made a call to a source in the Somerset County area, where that plane had apparently crashed. All my source said was "Get down here now!"

The whole way there we didn't say a word; somehow I think we knew that something really bad had just happened. We tried following some emergency vehicles to the site, but lost them. As we turned onto a dirt road in rural Somerset County we knew we were headed in the right direction, because people were standing outside their homes with their hands over their mouths. When they saw us they just pointed. We rounded a corner and could see an old strip mine in front of us. People were walking around like zombies.

About two hundred yards up we could see the emergency vehicles and smoke. We decided to drive closer, knowing at the same time not to get too close. I remember climbing out of the truck and nearly stumbling over a round piece of metal. I later learned it was part of Flight 93's engine. Keith started shooting, and I began looking for someone to talk to. About five minutes later a state trooper came running in our direction. Over the years you learn to work with the police while at the same time pushing the limit just enough to get what you need.

As I looked around I thought, there is no way this was a commuter plane. There was nothing but a hole, some smoke, and little pieces of metal everywhere. It had to be a small single-engine plane. Again, something inside told me to think again. The look on the trooper's face told us why. Breaking a reporter's "rule," we just climbed back in our truck and left.

With that angle gone, we had to find another way to get our video and sound bites. There was a road that basically circled the site like the rim of a bowl. At the very top we saw another one of our live trucks. The station had sent a second crew out, not knowing what had happened to us. Our reporter instincts kicked in. While one of us searched for witnesses the other crew got a live shot established. Within minutes we had a man who had watched the plane go down live on the air with our anchor back at the station. Minutes later, the other reporter and I were doing live shots, not knowing that the images we were sending out would be the first and only video taken of the initial devastation. Our camera was on the top of our mast shooting a huge crater and devastation for some two hundred more yards.

I think that's when the reality hit. Again another reporter's "rule" broken. Instead of worrying about my next cut-in or where I was going to get my information, my thoughts turned to my family. I thought of my four-month-old daughter, whom I had left for the first time since she was born, and my four-year-old son, who was in school just a few blocks from the Johnstown airport. I thought of my husband, who is part of Johnstown's special emergency response team and had been activated to search all incoming planes for bombs.

I kept thinking, "Oh, my God! What is happening and why?" Thirty-five years old and a mother of two, and all I could think about was calling home. With cell phones jammed I thought there was no way I could reach my mom all the way in Idaho. I quickly dialed her number, and it went through. When she answered I just cried. This story was something that every reporter dreams of covering. It was a huge national story, but at that moment I didn't want to. My mother reminded me that I wasn't doing my job for myself, but for others who needed to know what was happening. She told me she loved me and would stay by the phone. That was all I needed to hear.

Again the police pushed us back. We drove our live trucks higher, and in the next hour went on the air every ten minutes or so. Our IFB was the only connection to the world we had. They were feeding us as much information as they could get, which was very little. Our firsthand accounts of what we were seeing and several eyewitnesses were worth gold at that time.

All the initial reports were unconfirmed, and that's something you normally would never go on the air with. But this was different. At one point we were told there were 220 people on board Flight 93. There was also talk of a bomb and chemicals. Being careful not to scare people, we did the best we could to try to explain what they were looking at. About an hour later we again were told to leave—this time because it was now a crime scene.

About three hours had passed and by the time we got to the designated command post there were media everywhere. All my years of school, training, and field experience never prepared me for what I would go through during the next two weeks.

The first day was mayhem. Hundreds of reporters were scrambling for information. At press conferences it was hard to get a mic close enough to get sound. And I did so many live shots I was on autopilot.

After the 11:00 news that night was the first time we really had time to sit and reevaluate the whirlwind of events and what we needed to do. As a reporter you drool over stories where there are so many angles you could fill a hundred newscasts, and that's exactly what we had.

As media arrived from around the world we had to scramble to find eyewitnesses that hadn't already talked to fifty other reporters. And we did find them, including one eyewitness who watched the plane rock side to side before turning over and plunging nose first into the ground. He had been working at a strip mine that sat above the crash site. He said the plane was so low he could look in the windows. His eyewitness account helped investigators piece together what had happened just before the crash.

As day two rolled around so did the exhaustion. Not physical but mental. We agreed to work twenty-hour shifts that overlapped. Just enough time to go home, take a shower, and try to sleep for a few hours. As if that would ever happen. When I closed my eyes there was so much going on in my head I just couldn't sleep

I don't think any of us cared that we were tired and hot, or what we looked like for that matter. We were literally "live on the scene." It was grab a brush and lipstick if you could, but most of the time you couldn't. Living in cornfields for fourteen days changes some of the rules.

Eating or drinking wasn't a high priority. No one felt like eating, even though the community and businesses across the state had donated more food than anyone could eat in a year. Water was important, however, and we always tried to keep a full bottle nearby.

As the days passed the emotions grew. These were the kind of emotions that as a reporter and photographer you aren't supposed to show. It's hard to explain, but being "first" wasn't as important during those days as it is normally 365 days a year. Stations helped competing stations. Maybe it was because we spent every minute of the day together, talking about the same horrible tragedy. It wasn't unusual to see a photographer actually put his camera down and walk away from an interview because he couldn't take it. Veteran reporters would disappear into the port-a-potty and emerge thirty minutes later with red eyes. No one cared, other than to put their arms around them and give a hug.

I honestly think we put aside being "reporters," in the true sense of the word, and instead we were humans who had to experience and tell the world about stuff no one should have to hear. We had to watch the vans full of family members arrive, and look at the notes and mementos they left behind for their lost loved ones. We saw veteran police officers cry. And we walked away from scenes that were just too much. Reporter's "rules" didn't exist.

We also had to think twice about what we released. We had to remember what an impact September 11 was having on everyone. People were scared, and it would have been irresponsible to scare them anymore. Instead of telling people that all they found were teeth imbedded in trees, we would simply report that more remains were found. I really do think something happened to the reporters covering September 11. Empathy and sympathy became a key part of our reporting. To this day I think twice about what I say. Is a "juicy" piece of information that important?

For fourteen days straight we did live shots every hour, 90 percent without scripts or notes. We didn't need them. Every bit of information was burned into our minds and our hearts. I will also never forget the support we had out there everywhere we turned. That gave us all a good feeling.

But the only "peace" I really had, and what kept me going during that whole ordeal, was my family, and knowing that they were home waiting for me. I remember going home that first night to get a shower and seeing my children sleeping and wondering why all of this was happening and what was next. I grabbed several pictures of them and to this day keep them in my pocket.

In the field of broadcasting, you can get lost and obsessed with the job. I think anyone who has journalism in his or her blood knows what I mean. Not caring that you worked a sixty-hour week, and basically sacrificing everything to be "on air" or have an "exclusive."

I remember how much I fed on that adrenaline, but there comes a time when you must prioritize life. Jobs are great, but they aren't supposed to be your life. Someone once told me to always remember that your job may not always be there, but your family will.

About the thirteenth or fourteenth day of covering the crash of Flight 93 I had finally had enough. I wasn't sleeping, and I was having trouble eating. Our station had people for us to talk to; they too were concerned about what impact it was having. They brought in counselors that they had contacted just in case a reporter or photographer was having trouble dealing with what they saw and did. September 11 basically was uncharted territory for everyone and stations did not know what to expect.

I will probably feel the effects of covering this story for the rest of my life. To this day, I have trouble writing any story if it has to do with September 11. I find myself hesitating, not because I can't write it, but because it makes me remember. The flood of emotions always seems to come back. In fact, on the six-month anniversary of September 11, there was a big memorial at the crash site. I requested not to go. I just wasn't ready. I did cover the one-year anniversary, but I found myself having trouble sleeping in the days preceding. The night before I was in tears. But, one year later at the anniversary, being there and seeing everything good that has happened since that day really helped. I still would rather not cover stories at the crash site, and getting on a plane scares me to death. But I have learned to weed out the bad. Now I use all the good I saw from September 11 to make me a better and stronger reporter. Empathy, sympathy, and a realization that everything we say in a story affects someone, somewhere.

I remember all the stories I did in the following days of people coming forward to help out of the goodness of their hearts. I remember the small community of Shanksville, which lined their streets and homes with ribbons and flags out of respect for the victims' families. I remember the police officers who stood at attention in the heat for hours as family members visited the site. They did it not because they had to, but because they wanted to. What amazing stories. I now have a little more compassion for people, and I think twice when writing a story, knowing that what I'm saying affects people. I try to stop and put myself in the shoes of the people I am talking about.

A few weeks after September 11, I was still having trouble dealing with the events, and I remember questioning why I do this job. Why I choose day after day to tell people about all the bad and the tragedies in other people's lives. I questioned why I wanted to continue in this field and was almost at the point that I wanted to throw in the towel. That afternoon I came back to the station and had a million things to do before deadline. Stressed and worried I wouldn't get them done, I almost didn't answer my phone. It was the front desk. Someone was downstairs and wanted to see me. My initial reaction was to tell them to come back later, because I was too busy. But instead I told them I would be right down.

When I got to the front desk there was an elderly woman sitting there. She just smiled when she saw me. She told me her name and that she lived in Johnstown, and that she just wanted to see me and tell me thanks. That's when she handed me a little box. It was obviously tattered and torn from age. I didn't understand, but I took it and opened it. Inside was a small American flag made out of little red, white, and blue stones. When I looked up she was crying. I asked what was the matter and why she was giving this to me. She told me it was her way to say thanks. She said she no longer had any family living; her husband and children had all passed. She told me many years ago something horrible had happened in her life, and someone had given her this pin as a symbol of hope for the future. On September 11 she had no one to turn to and had never been so frightened in her life. She told me the only person who talked to her and told her what was happening and that everything would be okay, was me. If there is ever a day when I question why I chose to be a journalist, all I have to do is wear that pin, and remember.

Sherry Stalley is the co-anchor for the FOX 8 News in Johnstown, Pennsylvania. Prior to working with FOX 8, she was a crime and general assignment reporter and fill-in anchor for WJAC-TV in Johnstown. She has also been a reporter at KMVT News in Twin Falls, Idaho.

CHAPTER

12

The Serial Snipers

by Neal Augenstein, WTOP Radio, Washington, D.C.

The next bullet could come from anywhere, at any time. And, by the time the next victim realized he or she was a potential target while going about everyday tasks, it was too late.

That was the reality facing people living in the Maryland and Virginia suburbs of Washington, D.C., in October 2002. Two snipers, shooting long-distance, murdered ten people and wounded three others during what seemed like a never-ending twenty-two-day period. As residents tried to keep their families safe while maintaining some sort of normalcy, the region, the nation, and eventually the world became engrossed in the investigation of the person or persons who appeared to be randomly choosing targets while playing a deadly cat-and-mouse game, taunting police.

For reporters at WTOP-AM/FM, Washington D.C.'s all-news radio station, and all other news organizations, covering this story posed rare challenges for reporters, for a number of reasons.

No period of time felt "safe." Police scanners were running twenty-four hours a day. Citizens (and reporters) looked over their shoulders, wondering if a sniper was pointing a rifle at them. And the most daunting challenge: it appeared the killer was monitoring the media closely, perhaps fueled by the round-the-clock coverage, and could possibly use information disclosed in some reports to avoid capture and kill more victims.

One clear example of this: As the killings continued, police began instituting regional roadblocks immediately after receiving a 911 call of a shooting. Exits and entrances to major roads were shut down, and cars were painstakingly searched in an attempt to catch the killer in a dragnet.

WTOP's traffic reporters, who give round-the-clock descriptions of where traffic jams are occurring, faced a troubling reality: If they reported which roads were shut down, the killers could use that information to find another escape route. If they reported which roads were open, the fleeing murderers could use that fact, too. The solution was to report something along the lines of, "If you're stuck in backed-up traffic in Northern Virginia, it's because police are investigating." In other words, enough information to let listeners know *something* was going on, but not enough to compromise the investigation.

The physical act of covering this story presented some challenges. While the first five murders happened within a few miles of each other, the attacks spread to Washington, D.C., and Maryland and Virginia suburbs more than a hundred miles apart. Just getting to the crime scenes was stressful. With the police dragnets, it often took reporters hours to get to a crime scene. It's difficult to tell yourself, "Relax, you'll get there as soon as you can," when you know your editor expected you to be there "ten minutes ago."

Once a reporter makes it to a fresh crime scene, the first few hours are hectic. Between searching for a police supervisor or public information officer who can provide details, and trying to coordinate the logistics of feeding a report—everything from finding a place to park a satellite truck, to searching for a specific point where a cell phone signal works—there's not much time for comfort. Our reporters are told to keep a change of clothes in their work vehicle, as well as water, and some sort of nonperishable food. Some even carry an empty mayonnaise jar, for when a Porta-John or secluded tree isn't available. All of these things help when something like this event occurs and lasts for so many days.

After each shooting, briefings were held at Montgomery County Police headquarters. A virtual tent city was erected, although it took a while to get it right. With dozens of TV satellite trucks parked surrounding a parking lot, for several days, the asphyxiating fumes from the tail pipes were excruciating. Finally, the crews realized they should all park with their tail pipes facing out, which made things a lot more tolerable.

Well-funded news organizations rented phone lines so that radio reporters could use Comrex units, which substantially improve the sound quality of audio transmitted through a line. We even found that the plastic milk crates that carried extra-long cables could be used as makeshift tables when turned upside down. Most reporters and technicians brought lightweight foldable chairs for the hours spent waiting between the increasingly rare news briefings. As the days turned into weeks, a rhythm developed, and most people knew to stop for food before arriving at what became known informally as "the Moose Lodge," after Montgomery County police chief Charles Moose, who held most of the briefings.

It was very difficult for me to relax while the killer was still on the loose. My stereo and CD player gathered dust, as I didn't want to turn off my police scanner, for fear of missing anything important. At some points I'd be listening to the scanner while at the same time listening to my station and watching TV to see if other outlets had information we didn't have. Time spent on the computer wasn't for relaxation. It was for reading as many online newspapers and Web sites as possible to compare what they were reporting with the information WTOP had.

Thankfully, I was able to sleep well, and it did feel good at the end of the day to finally let myself (at least physically) relax. I made sure to get to sleep early enough, because I realized I was getting up too early every day, to read all the coverage I could find online, to try to understand as many angles of the story as possible. Since this story wasn't going away, I felt I had to come into the newsroom with ideas on how to advance it, even if there weren't any major developments in the investigation.

My diet suffered, too. I rarely prepare my own food at home, but during this story it simply never happened. I got into the habit of grabbing a fast-food breakfast on my way to work, missing lunch, and wolfing a fast-food dinner on my way home. I probably would have enjoyed a home-cooked meal, but I didn't feel like allowing myself the time to make one.

To be honest, it was a challenge to focus on anything that resembled a normal life. While I *did* make time to be with my friends, a lot of what we talked about involved the sniper story. Movies provided a few hours away from the near-constant attention to the scanner, the radio, and the television. I did try to walk around my neighborhood every night—just to get out and about, even though most of my thoughts were on the story.

Was I scared? The thought crossed my mind a number of times that if the sniper *really* wanted to gather media attention he'd attack during a news briefing, with dozens of cameras rolling. I can't say I sincerely feared for my life, either while working or just while going about my daily routine, but I did glance around when I filled up my car at a gas station. I tried to reassure myself by remembering a figure I'd heard, that there was a far greater chance of being killed in a car accident than being killed by the sniper. For some reason, that was comforting.

In between the periods of intense action, there were hours of inaction. I often had to report twice an hour even when nothing was happening. After writing my next report, I found myself trying to stretch often, to keep my body limber. Before reports I'd jump up and down in place, to get my blood pumping. I'd do the exaggerated stretching of my face and mouth, as well as repeating phrases like "Put a cup" to get my mouth warmed up. I drank a lot of bottled water and coffee. This was partly out of boredom, and in some cases the coffee was just to keep my hands warm.

The mental and emotional challenges of this story had some similarities to, and some differences from, other stories we cover on a daily basis. As any crime investigation develops, reporters and investigators often have differing agendas. Journalists want to report facts, while investigators want to avoid revealing many facts to protect public safety and eventually their prosecution.

During the initial stages of this shooting spree, Chief Moose held several briefings each day. He rarely provided any substantive information about what evidence had been gathered, or what progress was being made in the investigation. Most questions were answered with something along the lines of, "It would be inappropriate to discuss any portions of our ongoing investigation." This was frustrating for reporters, but probably good police work.

The fact that very little information was being shared at the podium should not come as a surprise. It's very rare for "news" to come from a news conference. The *real* news in any crime case comes from reporters "working their sources," holding private telephone conversations with well-placed people from any of the investigating agencies. These sources rarely are seen on TV or heard on radio, but at times they can be persuaded to share some information "on background" (which means you can't quote them by name) or "off the record" (which means you have to verify that information with another source before reporting any part of it). Of course, a reporter needs to remember that anybody willing to share private information isn't doing it "just to be nice." The person usually has an agenda that needs to be taken into consideration.

Despite the competition to broadcast stories and facts that other news organization didn't have, reporters in this case had to consider the consequences of their actions. At one point, a local TV station learned from a source that the killer left a tarot card near a shooting scene with the phrase "I Am God" written on it. What wasn't reported (and perhaps wasn't known by the TV station) until after Chief Moose angrily dressed down reporters for releasing evidence was that the card also included instructions to police to *not* provide details of the note to the media.

Ultimately, leaked information may have resulted in the capture of the man and youth arrested in the case. While Chief Moose announced from the podium the name of the man wanted for questioning in relation to the attacks, he chose not to describe the vehicle the man was thought to be driving. That information was leaked to news organizations, which broadcast the description of a 1990 Chevrolet Caprice and its New Jersey license plate number. In less than two hours, an alert driver spotted the car at a Maryland rest stop, and the two suspects were arrested while they were sleeping in the vehicle.

I don't believe I've suffered any post-stress events. I'm not sure if it's a good thing, but for some reason I sense I don't allow myself to feel the terror of horrific events. It's almost as though I say to myself, "You're being paid to be brave and keep working." A psychiatrist would probably have a technical name for it, but I feel like my job is to witness and document *other* people's emotions, and being a reporter is almost like my excuse to avoid feeling the intense emotions that I see others having. Maybe ten years from now I'll pay a price for that. I do talk to family and friends about "how I'm doing" during these types of stories, so that is an outlet.

After the arrests of the suspects, with the reign of terror ending, there was one more challenge. While the entire region was breathing a sigh of relief, I heard a temporary lessening of the journalistic requirement that a suspect be considered innocent until proven guilty. Obviously, reporters and anchors are human, and the belief that the horrible cloud of threat was clearing was enough to leave some journalists giddy. It reminded me that, while I am allowed to feel and express emotion, "it would be inappropriate," as Chief Moose might put it, for a reporter to do it too often while trying to provide unbiased reporting, even as this tragic story was finally ending.

Neal Augenstein is a reporter with the WTOP Radio Network in Washington, D.C., and is heard often on CBS News Radio. He got his start in radio news at UPI Radio Network, after years of producing radio talk shows.

CHAPTER

13

Sights, Sounds, and Smells of War—An Embedded Journalist in Iraq

by Ross W. Simpson, the Associated Press Radio Network

When I left home on February 27, 2003, the temperature was 28 degrees above zero and snowing. Almost three feet of snow covered the front lawn, and the local forecaster on the radio said the Washington, D.C., area could expect another ten inches before the blizzard subsided.

Thank goodness my daughter, Ginger, offered to take me to the airport to catch my flight to Kuwait. If it weren't for her Ford Expedition with its all-terrain tires and four-wheel drive, I would not have been able to get to the airport, much less to the Persian Gulf. My car, a 1989 Chevy Corvette, was buried in a snowbank.

As we pulled out of the driveway, a touch of nostalgia swept over me. I had left home many times during the past twenty-five years to cover conflicts and catastrophes around the world in places like Somalia, but this time felt different.

Unless Iraqi president Saddam Hussein had a change of heart and complied fully with United Nations resolutions to turn over supposed stockpiles of weapons of mass destruction, a "coalition of the willing," namely the United States and Britain, would destroy him and his army.

I had flown to Kuwait in February 1998 when it looked like the U.S. was going to war against Iraq, but Saddam had stepped back from the brink of war just before President Clinton pulled the trigger. From the moment I packed my boots, body armor, and battle fatigues for this trip to the Gulf, I had a gut feeling that there would be no brinkmanship this time.

Washington and Baghdad were on a collision course, and my wife, Judy, didn't have a warm, fuzzy feeling about this assignment. Neither did my four grown children. Even my four grandchildren held onto me for dear life when we said goodbye on the eve of my departure.

Seeing their sad little faces tugged at my heartstrings, and I couldn't help but feel guilty. Here I was at the top of my career with nothing to prove professionally. I had just won a coveted Edward R. Murrow award for my coverage of the terrorist attacks on September 11, but I was ready to roll the dice and risk my life to do something only a few civilians would ever get the chance to do: go to war with a Marine assault battalion. Sure, I believed what the U.S. was doing needed to be done, but I have to be honest—I just wanted one more chance to prove that at sixty I still had the "right stuff," whatever that is. The emotional roller coaster didn't end until I stepped off a 747 at Kuwait International Airport, where the temperature was 88 degrees, a 60 degree change in seventeen hours. The heat hit me like a ton of bricks.

The next two weeks were spent shaking off jet lag and getting prepared to join the First Marine Division in the Kuwait desert about thirty miles below the Iraq border. My old friend Jim Mattis, a lieutenant colonel when I first met him during Gulf War I, was now Major General James N. Mattis, commanding general of 1MarDiv. When we spoke before he left Camp Pendleton, California, in early February, he asked where I wanted to be in his division. I replied, "Where you would be. On the tip of the spear."

When the division public affairs officer, Captain Joe Plenzler, announced the "embed" assignments on March 11 at the Hilton Hotel in Kuwait City, I learned that I was going to First Battalion, Fifth Marines, the most decorated battalion in the Marine Corps.

Correspondents like me were willing guinea pigs. We had begged for years for permission to accompany front-line forces into battle, and now our dream was about to turn into a nightmare.

Before I left the Hilton, I ran into an old friend from NBC. It would be the last time I would see David Bloom. He died of a pulmonary embolism from sitting motionless in his armored vehicle for long periods of time. I felt his pain. My legs ached so badly I almost screamed. Sometimes we would be cooped up in close quarters for hours on end. We got out long enough to relieve ourselves in a ditch alongside a highway, never long enough to restore circulation. I'm surprised more of us didn't meet similar fates. I didn't know Michael Kelly, editor of the *Atlantic Monthly*, who would also die in the days ahead, but he stood next to me as we were interviewed for ABC's *Nightline*. In all, thirteen members of the media, men and women, died covering Operation Iraqi Freedom. A heavy price, percentage-wise, when you consider the number of embedded correspondents and camera crews.

With the exception of NBC (Nuclear, Biological, and Chemical) training at DoD-sponsored "media boot camps" in the United States, nothing prepared me or my colleagues for the combat that we experienced "up close and personal."

Survival was the name of the game during the war. Being run over by an armored vehicle was my greatest fear. At times, I felt like Charlie McCarthy, the puppet whose head could spin 360 degrees.

The first couple of weeks in the desert were downright miserable. Temperatures soared into the 90s during the day and plunged into the 30s at night. Then there were the sandstorms and the thunderstorms. It was impossible to sleep with torrents of rain pouring in the open gun turrets. By morning, the column looked like it had been dipped in a vat of milk chocolate. Mud coated the guns and belts of ammunition. I can remember going days without sleep.

I weighed 265 pounds when I stepped off the plane in Kuwait. When I boarded a flight home fifty-two days later, my weight had dropped to 228 pounds. My blood pressure had also gone down from 156 over 96 to 120 over 70, the best BP in more than forty years. I attribute the weight loss to exercise and lack of good food.

MREs are truly "meals rejected by everybody," but they were gourmet meals compared to what the Iraqi soldiers were eating. MREs were hot. That's about all you can say about them. If you weren't present when the platoon sergeant dropped off a case of rations, you got what your buddies didn't want: Pasta Alfredo and other vegetarian meals.

There were four members of the media assigned to 1/5. All of us in the autumn of life. No wonder the battalion commander, Lieutenant Colonel Fred Padilla, wrote the following observation in his personal journal on March 12: "We received our embedded correspondents today. Representatives from AP, UPI and two guys from the Orange County Register. . . . They are old and out of shape, and I fear they may not be able to keep up. They are really feeling the heat." We must have looked like a sorry lot when "Hondo" welcomed us to the "Fighting Fifth." But Rick Tomkins from UPI and the Orange County Register guys, photographer Mark Avery and writer Gordon Dillow, a Vietnam veteran, proved Padilla wrong. We kept up all the way to Baghdad with young Marines a third of our age.

Going into combat is a real rush. The sights, sounds, and smells will last a lifetime. Especially the sounds of artillery and aircraft softening up enemy positions with hundreds of high-explosive shells and bombs. Add in the sight of burning oil wells and the smell of thick smoke, and you can begin to imagine what it was like to be with the first U.S. forces that crossed into southern Iraq. The battalion was ordered to cross the "Line of Departure" a full nine hours ahead of schedule when plumes of smoke appeared on the horizon, a sign that the Iraqis were torching the Rumaylah oil fields and pumping station.

The weather was rotten when 1/5 crossed through two breaches in the ten-foot-high sand berm that separates Iraq from Kuwait. Visibility was down to a few hundred feet. The Amphibious Assault Vehicles (AAVs) were turning dirt tracks into dust bowls, a foot deep in places.

Fifteen seconds after 1/5 crossed into Iraq, the battalion radio crackled with an immediate call for help from Alpha Company just ahead of us in the darkness. "Alpha is engaged with three Iraqi tanks and dismounted infantry," radioed the young company commander as he called for one of our Javelin teams to get out of his AAV and take out the tank. Counter Mech, my platoon, was ordered to come up and provide covering fire. We could hear the gunfire and see tracers bouncing off the ground, but there was no time to be scared. The radio was alive with transmissions as Counter Mech moved in for the kill.

I could hear the commander yelling on the squad frequency for the two-man Javelin team to "take the shot." They did, and the tank exploded, incinerating its three-man crew.

More than one hundred enemy soldiers were killed in and around the pumping station on the first night of the war. I saw one of them the next morning. The army captain looked like he was sleeping on the edge of a tank revetment. As I knelt down beside him, I could see that the back of his head has been blown away by a sniper's bullet during the night. I thought to myself, what a waste. He was someone's son, husband, and father, but no one would ever know what happened to him. His men had either been killed or had run away, leaving his corpse to rot on the ground or be eaten by wild animals. The Marines I was with didn't take time to bury him or other Iraqis they killed ten days later at the Saddam Hussein Canal in south-central Iraq. They were just chunks of meat. "Better him than me," said a Navy corpsman assigned to my platoon as he watched the left foot of a dead officer's leg burn in the flaming wreckage of his motorcycle. I must admit I felt the same way. It's hard to have any feelings for someone who would have killed you if he'd had the chance.

When Counter Mech rolled up to the canal, Gun-1, the HUMVEE I was riding in, came under immediate fire from a ZSU 23-4, a four-barreled antiaircraft battery dug in about four hundred meters down the waterway. My vehicle commander ordered everyone except the gunner to get out of the vehicle and seek cover behind a small building. But I froze. For reasons I can't explain, I sat in my seat and watched as Lance Corporal Dominic Chevalier, three days shy of his twentieth birthday, smothered the enemy battery with hundreds of .50 caliber slugs. I really don't remember getting out of the Hummer, but I do remember at least four artillery and mortar rounds exploding in our midst. A 122 mm enemy mortar round hit a few feet behind my Hummer as I panned my video camera toward the bridge. "*Wow!*" was all I could say as the vehicle disappeared in a cloud of dust. On the videotape you can hear the gunner and the vehicle commander, Corporal Jeremy Mahon, yelling, "Ross, get in; we're going."

Our news organizations constantly reminded us that we were noncombatants. We were not supposed to assist the military in any way, much less pick up a weapon and fire it, but during a live broadcast to the nation's news directors in Las Vegas I reminded them that there is no such thing as a noncombatant on the battlefield. The only rule is "Kill or be killed."

After silencing the AAA battery, Counter Mech crossed the canal and engaged other Iraqis dug in on both sides of the road. I watched as three Marines walking beside my HUMVEE suddenly swung their weapons toward something moving in the underbrush. I watched in horror as they machine-gunned a soldier who was desperately trying to reach his AK-47. The enemy soldier's body jerked like a puppet on a string as 5.56 mm slugs ripped through his legs and arms and blew away his lower jaw in a macabre mist of blood and bone. Tracer bullets from Corporal Warren Rutledge's squad automatic weapon ignited the gas tank on the officer's motorbike. The fire burned for more than an hour before Corporal Travis Brozowski, a member of my platoon who scored the first TOW missile kill of the war, got a fire extinguisher and put out the flames. I watched as he stepped over the dead soldier as if stepping over a piece of wood. Brozowski never looked down. He focused on the fire. The date was April 1, but the only fool was the Iraqi who resisted. Had he thrown up his hands, he would be alive today.

Hundreds of Iraqi soldiers died in air attacks during the twenty-one-day war. I will never forget the sweet pungent smell of roasted human flesh as 1/5 crossed a bridge over the Euphrates River south of Baghdad. It was about midnight and some of the shells in the enemy vehicles were exploding as the flames reached the ammo chests. The smell of death hung heavy in the night air. It filled your nostrils and made you sick to your stomach. The closer we got to Baghdad, the tougher the resistance, culminating in what I referred to in one of my live battlefield reports as the "longest drive-by shooting" in history.

The battle began at three o'clock in the morning on April 10 as 1/5 entered the north side of Baghdad and headed for Azimiyah, one of Saddam Hussein's favorite palaces, on the Tigris River. More than one thousand rocket-propelled grenades were fired at the battalion between 3 A.M. and 6 A.M. as it wound its way through narrow city streets like a giant steel serpent. By the time the armored column arrived at the front gate of the palace, an RPG had hit every track at least once. Some had been hit several times. Two of the deadly shoulder-fired missiles flew across the hood of my HUMVEE and exploded against the track ahead of us.

Three Marines were wounded and Rick Tomkins was shaken up. The driver of a HUMVEE behind me saw a Fedayeen guerilla in black pajamas fire an RPG at my side of the vehicle. It sailed over the hatchback behind my head and hit a building, breaking in half. A dud. Later, in a traffic circle where Alpha Company's gunnery sergeant was killed, another guerilla fired an RPG that went under my seat. What can I say? It just wasn't my time to die.

The battle for the palace lasted late into the night, before Marine air broke the back of a planned counterattack.

After posing for a picture with seventy-two friends in the rubble of the palace, I left Counter Mech Platoon on the afternoon of April 16, stopping briefly at General Mattis's headquarters in the building once occupied by Saddam's secret police, before catching a helicopter that took me to a forward refueling base on an abandoned Iraqi airbase, where a Marine cargo plane picked me up and took me to Kuwait.

As I thanked Mattis for an incredible experience, he simply nodded and said, "Get a shower and go home."

I showered at the Sheraton Kuwait on March 10 before heading to Camp Commando in the Kuwait desert. The next time I showered was on April 17 at the Crowne Plaza in Kuwait City.

The young night clerk stared at me when I walked up to the registration desk and told her I needed a room. "You need more than a room," she giggled.

I can't begin to tell you how good that warm water felt. Had I not been so tired, and fearful of falling asleep and drowning, I would have been contented to spend my first night back in the world in that tub.

Since the war I have spoken to groups across the country about my experience as an "embedded correspondent," but if I live to be a hundred years old I will never forget the "Sights, Sounds and Smells of War." They are intoxicating, and I just might be tempted to try it again.

Ross W. Simpson has spent forty-two years in broadcasting. Currently he is the morning anchorman on the Associated Press Radio Network. Ross has won many awards in his profession, including an Edward R. Murrow Award for his live coverage of the tragic events of September 11. His live coverage of the war in Iraq earned him the 2003 Gramling Award from the Associated Press.

CHAPTER

14

Stories That Haunt You

by Desirée Berenguer, News 12 New Jersey

My cameraman Joe and I have been working together for three years at News 12 New Jersey. We are a team that works well together, but we are also great friends. So many times we ask each other, can we go through this again? Feeling the stress of covering a crisis sometimes makes us feel like we are in hell. You can get very burned out in this business. You constantly have to ask yourself, why am I going through this daily grind? You have to look for the positive things that come out of crises, but first you have to experience them.

It was January 1998; I was working at WNDU-TV, an NBC affiliate in South Bend, Indiana. I had been there for about four months as a general assignment reporter. Having been raised in Florida I still was not used to the cold Midwest weather, even though just months before I had lived in Maryland. I will never forget all the days we had of "lake-effect snow." It was so cold your eyelashes would freeze!

On Sunday, January 19, the weekend assignment desk called me in early because of a fire in Elkhart, Indiana. This was about thirty minutes north of the station. My photographer that day was Russ Govert. We quickly geared up and headed out. What I saw next will haunt me for the rest of my life. We watched in horror as Elkhart paramedics pulled body bag after body bag from the dark-green two-story house at 919 West Garfield Street. We stood there for a moment in silence trying to fight back the tears. I had covered fires in the past but this one was different. . . . Children's lives were lost.

Firefighters say the flames engulfed the house in a matter of moments; all that was left were charred remains. Six children and one adult died on that tragic day. Investigators say it was the worst fatal fire in Elkhart's recent history.

It was difficult to get into reporter mode. We are trained as reporters not to show emotion but just to report the story. In a few hours I had to put myself together emotionally and physically because we were going live for the 6 P.M. newscast. But I just could not get the images out of my head. Six children between the ages of five and nine had passed away, along with their neighbor, sixty-five-year-old Charlie Lemons. He escaped the fire then went back in to rescue the kids and was killed.

As I was running around on that cold day, I remember the sun was shining but ice was all over the sidewalk. The neighborhood was on the Elkhart's south side. It was run-down. Firefighters were still putting out hotspots in the house. Ice had formed from all the water from the hoses. The temperatures were so low I could barely move my fingers to write. I was trying to get a statement from Tony Johnson, Elkhart's fire chief. In the meantime, I talked to neighbors who said they heard someone screaming, "My babies . . . my babies are inside . . . help them." They told me Nettie Sims lived inside the house with her grandchildren. It was unclear how many people lived there. But one thing was for sure: Nettie Sims made it out alive. She was rushed to Fort Wayne Burn unit.

The next day Fire Chief Johnson had a press conference. He told reporters a makeshift heating device rigged from a stove burner started the Sunday-morning fire that killed seven people. The gas had been shut off in the two-story house since July. I asked the chief if a smoke detector was found. He said at this point one had not been found. A working space heater was also found in an upstairs room but the device had not caused the fire. Then another reporter asked about the children's mother. Three of the six kids belonged to Earline Hawkins. She had left the house just a few minutes before the fire started. Prosecutors say Nettie Sims, the grandmother of the children, admitted to using cocaine the night of the fire. They alleged that cocaine use on the part of Sims and her daughter Earline led to the neglect of the children. The story was just getting worse.

A few days later the children were buried. It was so sad to see the small caskets lined up in front of the altar of the church. Sitting on top of the closed caskets were picture of the kids. They had such beautiful smiles. I didn't know them personally, but they looked like they were full of energy. Their elementary teachers told reporters that, although their home life was not the best, they were wonderful kids in school. Teddy bears and small bouquets of flowers filled the church. As you can imagine, the family, the entire community was distraught. The church was full with almost a thousand mourners. Everyone was overcome with tears. One woman read letters from the children's classmates. "You did not even make it to the third grade. I'm sorry you did not get saved by the hero." Another wrote, "I love you. You are my best friend. You are in a better place."

I haven't been back to Elkhart, Indiana, since June 1999. I left one month before Nettie Sims's trial, but I follow the story even today. A few months later, Nettie Sims was released from the Indiana Women's Prison. She had been convicted of neglect for the 1998 fire that killed six of her grandchildren and her neighbor. The plea agreement offered in court was a ten-year sentence with seven years suspended. Counting time served prior to conviction, authorities released her after she served half of the three-year term.

Sims's daughter, Earline Hawkins, remains in prison, serving a fifteen-year term for charges related to the fire. Hawkins lived with her mother on Garfield Avenue. We found out later that Earline was the woman outside screaming, "My babies . . . my babies are inside." Both women will never be the same.

The flames of the Garfield Avenue fire are still in my consciousness. The images of the seven body bags are still planted in my head. Any winter that I'm sent out to a story on home heating problems my mind rewinds to the conclusion of investigators that a makeshift heater sparked the tragic fire. Every time I cover a fire I think of those sweet, innocent souls who were lost on that day. I say a quick prayer every time we arrive on a scene hoping that no lives were lost. But I constantly have to remind myself I am here to do a job, not to get involved emotionally.

Fire stories like this one are not the only stories that haunt me. I've covered many traffic accidents over the years, but there was one that I'll never forget.

It was May 1, 2003. I had been the morning reporter at News 12 New Jersey for almost three years. I arrived at work around 4:30 A.M. Normally I am sent on a feature live shot that is booked in advance. On that day, my live shot had been canceled. My producer said a car accident just came over the pager and that I should head over to Linden in Union County. We normally don't cover car accidents but this apparently was a disastrous collision that closed down a major highway in New Jersey.

My photographer, Joe Caruso, and the satellite truck operator, Rick Sullivan, packed up, and we all headed out. Linden isn't too far from the station. About twenty minutes later we arrived. Routes 1 and 9 northbound and southbound were closed for about a mile, which was going to be a traffic nightmare for the morning commute.

The area was blocked off with police tape. All the other media stations were already set up to go live for the morning newscasts. As we were walking to the scene of the accident, Joe and I both saw the same frightening thing. Body parts were hanging out of the cars. Both cars were indescribable, with mangled pieces of metal scattered everywhere. Police say Edward Belvin, an off-duty New Jersey cop, was driving his Mercedes sports car southbound on Routes 1 and 9. He crossed three highway lanes, jumped a divider, and collided head on with a Ford Taurus. Police say the Taurus was carrying five immigrant workers who were heading home. Ironically, the collision happened right in front of a cemetery. The cars were torn to pieces by the force of the crash.

No matter how many times I had seen dead bodies, this still gave me a sick feeling in my stomach. How could this happen?

Six people died on that day. The workers, who lived in Elizabeth, were employed by a Houlihan's restaurant in Holmdel, which they had left about an hour earlier. Police say that Belvin was driving drunk. His blood alcohol level was three times the legal limit in New Jersey. Sadly, the blood test results only served to intensify the tragedy that took place that morning.

I don't understand how police officers and rescue crews can see this sort of tragedy every day. I know it's part of their job, but I still get queasy when I see blood and body parts, and just remember the horror of that scene.

It's weird how certain stories stick with you. I didn't know this immigrant family but from interviewing their employer I learned that they came here for the American dream. All of them were under the age of thirty-five. The police officer who crashed into them was thirty-five years old. Lives lost for no reason. To this day, every time we drive by the scene Joe and I become silent. It's an eerie feeling.

For all the stories that haunt us there are so many wonderful stories that keep us going back for more. So I think of the positive stories that can make a difference in our community. This gives me some relief from the stresses of this job.

It brings me to my next story that will always stay with me, which shows me the positive ways of the world. I guess I should say one particular person will always be in my mind. Tom O'Leary is the executive director of Samaritan Homeless Interim Program (SHIP), a nonprofit organization in Somerset County, New Jersey, that feeds the homeless.

Joe and I first met Tom three years ago during Christmastime. Every holiday season each reporter at News 12 New Jersey has to choose one person or a place that is helping the community. We call them our "holiday heroes." We chose Tom. Tom can understand what the people who come to his soup kitchen are going through. Tom was once homeless, jobless, and a drug addict. He turned his life around and now has helped thousands of people get back on their feet as well.

We went to one of his soup kitchens at a local church, where there were hundreds of people in line for food. People of all ages, even kids, were there. All of them had a desperate look in their eyes and wanted the same thing. It was so sad to see, but it made us realize what we have and how lucky we are.

Tom told us the story of a lady with three children. She took the meal and started crying. She confided that she had been on her way over to the shopping center to steal food.

SHIP also provides mobile soup kitchens. During the cold winter months is when you see the most people come out. Every Thanksgiving Day we do a live shot with Tom and all his volunteers at the Time-to-Eat Diner. They cook hundreds of turkeys with all the trimmings so the folks who come to his soup kitchen get a nice hot holiday meal. Unfortunately, in the past three years the amount of food being cooked is getting bigger. The startling statistics are that the number of homeless people in New Jersey is going up. The cost of living is so expensive it leaves many people hungry and without shelter.

Tom had definitely found his calling; he has left such an impression with Joe and me. He is on a spiritual and moral mission to reach out to those in need. Because of what he does, he gives me the strength to go out and do my job. I always try to report on a story in a fair and balanced manner. I try to tell both sides of the story and remind viewers of what's going on in our world whether it's good or bad. The stories that haunt me from the past help me keep this focus.

Desirée Berenguer is the morning reporter at News 12 New Jersey. She also fills in as an anchor on the morning and afternoon editions. Prior to coming to News 12 New Jersey, Desirée worked at WNDU-TV, an NBC affiliate in South Bend, Indiana, where she was a general assignment reporter and fill-in anchor. Before that, Desirée worked at WHAG-TV, an NBC affiliate in Hagerstown, Maryland. Desirée has been the recipient of several awards for best feature stories from the Associated Press, North Jersey Press Club, and the Working Press Association.

ON THE FRONT LINES WHEN SEVERE WEATHER STRIKES

"The reporter . . . is the person standing in the rain and wind and amidst the wreckage, but that is not an excuse for the rapid and breathless delivery that is commonplace on television today. . . . Field reporters must sound like they are in control of themselves and the situation."

Elliott Wiser

"Remind your staff . . . who will be in the field during the winter storm, to slow down. . . . The adrenaline rush that generally helps them be such good broadcast journalists can be hazardous to their health. . . . No story is so important that they can't take just a moment to make sure they're being safe."

Dan Shelley

CHAPTER

15

Hurricanes and Tornadoes

by Elliott Wiser, Bay News 9, Tampa Bay

The bright blue sky had turned a light gray. The palm trees began a familiar wind-induced tilt. I looked out my office window and there it was. An ominous dark gray, almost black, cloudbank advancing from the southwest. It was September 1998 and Hurricane Georges was bearing down on Tampa Bay. This was not a surprise, because for almost a full day Bay News 9, the area's twenty-four-hour news channel, had been going wall-to-wall with hurricane news and weather.

The storm never did make landfall; it veered northwest at the last minute. There were some high winds, crashing surf, and heavy rains. When the skies finally cleared again, Bay News 9 had been on the story for almost forty-eight hours. Some of us stayed the whole time, while many were able to go home and get some sleep. Yet all of us were placed under a tremendous amount of stress. It never showed on the air.

And that is the key to covering severe weather. It is like that deodorant commercial: the audience should never see you sweat. Reporters and anchors need to remain calm and in control, no matter what they are thinking inside. This chapter will offer practical advice to everyone who has to cover hurricanes, tornadoes, severe thunderstorms, flooding, and earthquakes. (In the next chapter, Dan Shelley will have some excellent advice for you on covering blizzards.)

Just remember, whatever you do, stay calm and don't let them see you sweat.

HAVE A PLAN

The station with a breaking news plan is the station that wins breaking news coverage. If you are an anchor or reporter, ask the news director about the plan, and if

In the fall of 2004, within the span of six weeks, Elliott and the Bay News 9 staff prepared to cover four major hurricanes. Three of those storms struck the Bay News 9 coverage area. Elliott reports that the information contained in this chapter proved correct and survived its "battle test."

there is not a plan then volunteer to write one with the input of news and operations employees. If there is a plan, then make sure everyone has a copy. Talent responds to a well-designed plan. The newsroom will be chaotic, but if everyone knows what he or she should be doing then anchors and reporters will feel like part of a news-producing machine as opposed to witnessing mass chaos. Before examining what reporters and anchors can do to improve their performance in severe weather, it is important to prepare the newsroom for climatic events.

Here is what a station's breaking news plan should contain:

- A list of every employee and a way to contact everyone. Make it clear whether people should call or just come to the station.
- Have your general manager agree that everyone in the station should help when there is severe weather coverage. You cannot do it alone. If your general manager does not agree with this, then your station is not ready to win severe weather coverage.
- The general manager, sales manager, and news director should talk beforehand about how long coverage should last. There is nothing worse than watching an argument unfold during an earthquake or hurricane. If your station is not prepared to do wall-to-wall coverage during a major event then communicate this beforehand to key people. Your staff will not like watching someone else going wall-to-wall while you are running regular programming. Nothing kills a performance quicker than a noncompetitive news station.
- Designate who should be in the booth and who should talk to the anchors. Limit the latter to an experienced member of your news management team.
- Restrict access to the anchors. Designate one runner to deliver copy and other pertinent material to the talent.
- Have one person communicate with the crews, so reporters are not given conflicting instructions. Designate a runner between the assignment desk and the control room so the control room coordinator knows who is going where.
- Designate a few administrative folks (human resource and business managers are perfect for this) to handle provisions. You should have extra drinking water, nonperishable food, flashlights, batteries, rain gear, tires, cots, blankets, towels, antibacterial moist towelettes (a shower isn't always available), and petty cash.
- Make arrangements with a nearby hotel or motel to provide rooms in case of a long event.
- Contact local restaurants to deliver food.
- Make sure you have a backup generator.

There are many more things that should be done to prepare for breaking news, but if you meet the above criteria then the newsroom has made a good first step toward providing a controlled atmosphere that will allow the talent to succeed. Now that the station is ready, let us prepare the talent for severe weather coverage.

GETTING READY FOR A STORM

The anchor or reporter should prepare for severe weather on those days when there's a clear blue sky and no threat of storm. First, make sure your family knows where to go, whom to call, what to do in case of severe weather. You need to develop an emergency plan at home so you can concentrate all of your energy on your work. Then start packing.

Find a waterproof bag and fill it with a change of clothes, boots, toiletries, extra makeup, hair products, medicines, glasses (in case you lose a contact lens), water bottles, snacks, and money. That bag should be placed in your car or station locker. Again, being prepared will allow you to concentrate on your job.

"I keep one under my desk," says Bay News 9 senior reporter Chris O'Connell. "My bag consists of clothes for two days, an extra pair of comfortable shoes, toiletries, two towels, and my rain slicker and rubber boots. I also carry a small thermometer, anemometer [wind speed indicator], and rain gauge for on-air demonstrations of the weather at my location."

Reporters and anchors should also do their homework. You should develop a severe weather notebook that contains facts about past major storms in your area, a list of evacuation centers and routes, basic weather terminology, a list of the area's emergency officials, and viewer-friendly tips.

"One thing I do is jot down several weather terms," recommends Bay News 9 morning anchor Jen Holloway. "I promise you will have a better chance of sounding knowledgeable if you know what you are talking about."

Finally, you should develop a routine that allows you to "calm down" before you go on the air. Whether it's breathing exercises, meditation, or visualization, remaining calm in the eye of the storm is essential. Practice this before the storm hits. Do not fool yourself into thinking that when you're in the middle of bad weather your natural instincts will take over and you will stay cool. This is something that should be practiced, and practiced often.

IN THE EYE OF THE STORM

Now let us consider performance enhancement techniques during severe weather for three job categories: the news anchor, the weather anchor, and the field reporter. There are similarities in what constitutes a "good" performance, but each job brings its own unique challenges.

News Anchor

The news anchor is faced with the prospect of keeping the coverage coherent and timely. The best anchors act like traffic cops who calmly direct the audience to the next new development.

"We are constantly being fed information from the desk and my job is to make sure the material is clearly understood," says veteran anchor Al Ruechel. " If we don't know something we tell the audience."

Ruechel makes an important point. Anchors need to be honest with viewers and tell them everything they know and do not know, especially during the preliminary stages of severe weather coverage. The viewer can connect with the anchor who asks the same questions the audience is asking. I think it is important for anchors to point out that "we don't know the answers to these questions, but we have crews on the scene who will find out." This is where note-taking plays a key role.

I coach anchors to divide their paper in half, or to keep two separate sheets of paper. On one, the anchor should write the key facts of the story. This helps when summarizing the story (which should be done often during severe weather coverage). This is your basic Who, What, When, Where, and Why. On the other sheet, or half-sheet, the anchor should write down the latest information. The information on this sheet of paper is constantly updated, because the anchor should always highlight the latest information. During hurricanes it is important to note road closures and emergency situations. The anchor who can say "This just in" or "We have a developing story" will make the station's coverage sound fresh. Note-taking can assist the anchor in staying focused and highlighting the latest information.

The information is important, but so is the way the information is conveyed to the viewers. Ruechel adds, " I also make sure to remain calm and reassuring and make sure all things are taken in their real context." An emotional anchor does the audience a great disservice during severe weather. I have seen anchors talk about tornadoes with that gleam in their eye. Sure they are excited about doing breaking news, but someone has just lost his or her life or property. I think anchors need to be considerate and sensitive when delivering severe weather coverage.

Anchors should avoid talking about their personal situations. The last thing the audience needs to see and hear is an anchor complaining about his or her problems. Viewers do not care about the anchor's troubles; they have their own worries. Rather than the anchor saying, "I've driven through the rain and it's tough to see," the better way to connect with the audience is to say, "You may have problems seeing the car in front of you when driving outside, so be extra careful."

Weather Anchor

The weather anchor faces many of the same challenges as the news anchor, but during severe weather the weather anchor needs to have more answers than questions. Also, note-taking is more difficult for the person standing in front of the chroma key. But the weather anchor needs to highlight the latest information and think of the viewer's needs.

Alan Winfield, the chief meteorologist at Bay News 9, says, "All format goes out the window during severe weather. I try to showcase what is most urgently needed to be seen by the viewer." The weather anchor needs to decide if weather franchises are to be included in severe weather coverage, but that discussion should occur well before a storm hits. The weather anchor should work hard to highlight the latest information and then provide historical context. For instance, rather than beginning a forecast with "A severe thunderstorm warning was announced three hours ago," the anchor should say, "We have heavy rains and high winds over Metropolis and moving east," etc. The weather anchor should also showcase the latest weather technology.

Reporter

The reporter has the most difficult time trying to remain calm and composed. This is the person standing in the rain and wind and amidst the wreckage, but that is not an excuse for the rapid and breathless delivery that is commonplace on television today. I am not advocating that the reporter deliver the story in a monotone voice, but this is where relaxation exercises are a must, because field reporters must sound like they are in control of themselves and the situation.

Reporter Chris O'Connell offers even more important advice: be safe. "When a storm hits, it's our job to get the story but it is also easy to lose sight of personal safety. When roofs are flying off houses or lightning is roaring overhead remember to ask yourself, 'Am I safe?' Don't get caught up in the excitement of severe weather. Show viewers the reason why you are out in the elements, but keep yourself safe."

The most common question reporters like O'Connell are asked after severe weather coverage is why they are standing in the middle of the storm telling people not to go outside. If the weather is so bad, why are the reporters outside? A good reporter will answer that question while he or she is live.

The reporter covering severe weather also has another concern—what to wear. A big hood that covers one's face is a bad choice for obvious reasons, and not wearing any outerwear makes you look like you really are too dumb to come in from the rain. The answer is station raingear with a station hat. Some reporters can get away with other types of clothing, but I would play it safe. If your station does not have foul weather gear, then talk to your news director and marketing director.

The good reporter is able to describe the scene with a strong command of the language and an unparalleled storytelling ability. The best reporters not only describe what they *see,* but also what they *smell, hear, and touch.* Sensory descriptions are a powerful storytelling tool too often ignored by television news reporters.

Other Advice

Finally, here are some tips about your care and feeding during severe weather coverage:

> When you're in breaking news for hours it's not easy to sit in the same chair for hours on end, so try and stretch during a package or weather report. Just stand up and stretch—there is no need to unplug your

> microphone or IFB. As for eating . . . avoid chocolate, milk products, and thick liquids. These products coat your throat and you can't clear it. Avoid soda. Instead, look for warm liquids.
>
> Jen Holloway

> I find I pace myself and try to get into a rhythm. If you are a runner you know what I mean. When you run, the first mile or so you are setting your pace and trying to get into the zone. Once you are in that zone it seems that you can run forever. During severe weather the same thing happens and that becomes obvious to the control room; they can almost figure out where you are headed next in your weathercast.
>
> Alan Winfield

This is great advice from people who have spent hundreds of hours doing continuous news coverage during severe weather.

CLEARING SKIES

The severe weather has come and gone, so how do you relax and recover? First and foremost, avoid drinking too much alcohol when you get home. The same goes for anything that will disrupt your sleep pattern. Chances are your body has already been thrown out of kilter by the long hours you just worked. The best plan is to go home and not try to immediately go to sleep. The adrenaline is still pumping through your body and it will take a few hours for you to calm down. The best suggestion is to read a book, play a computer game, or watch television (not news). The only way your body will recover is to remove yourself from stressful stimuli.

Protecting your voice is also important. Clear, warm liquids are best. Also, this may not be the best time to begin long telephone conversations. You need to allow your voice to recover.

Severe weather coverage is the cornerstone of any successful station's blueprint for success. The station that wins is the one that features experienced reporters, anchors, and meteorologists who calmly communicate the latest information. Some practice and a good severe weather plan are the best ways to prepare for the day when the blue sky turns gray.

Elliott Wiser is the vice president/general manager of Bay News 9 and vice president of news programming for Bright House Networks. He is the former vice president of news programming for Time Warner Cable. Elliott was also part of the launch team for Northwest Cable News in Seattle. A former news director and reporter/anchor at the CBS affiliate in Richmond, Virginia, Elliott has covered everything from hurricanes to earthquakes.

CHAPTER

16

Severe Winter Weather

by Dan Shelley, WTMJ Radio, Milwaukee

During the quarter century that I've been in this business, I've covered my share of blizzards, ice storms, and other harsh winter weather. I've had my share of close calls. I've gotten my share of battle scars.

There was the time I narrowly escaped injury, or even death, when my news car careened down an icy embankment. It was during an ice storm that was so bad you had to wear spikes on your shoes, à la a utility pole worker, just to walk a few feet. I was driving to the scene of a bad ice-related crash when a police car pulled up behind me, red lights flashing and siren wailing. When I tried to pull over to let the squad car pass, I kept on going, right down the embankment. I was shaken, but, thankfully, not stirred to the point of being hurt.

Then there was the time I did serious damage to my leg when I slipped on the ice as I ran out of the station to cover a story. It was after a blizzard that dumped eighteen inches of snow. I don't even remember what the story was. But I certainly remember bounding out the door, then promptly cracking my leg on some ice-covered concrete stairs. I ultimately had to have surgery and missed several days of work.

And then there was the time I got sunburned because the only way to avoid frostbite was to stand right next to the flames while covering a fire. It was the middle of the night. The snow was falling. The wind was howling. It was below zero. The burning farmhouse was already a loss when the rural fire department arrived, so all the firefighters could do was stand near the flames to keep warm. A crusty fire marshal from whom I needed information urged me to do the same. While I was interviewing him, a curious neighbor wandered up. It was a teenage boy who, despite the snow and the wind and the bitter cold, was wearing nothing but his skivvies.

"Hey," the kid chortled, "what's going on?"

The crusty fire marshal sneered back: "What are you, the product of inbreeding?"

I must admit that I have never been accused of being the product of inbreeding, but I have learned my share of lessons, mostly the hard way.

In all the time I was a field reporter, I gave nary a thought to anything other than pursuing the story. The adrenaline rush made me think I was immune to the inherent hazards associated with doing my job. Quite honestly, it never even occurred to me that there might be inherent hazards.

I'm smarter now. These days I stay inside the station. Still, that approach to covering a winter storm has its own hazards, not the least of which is sleep deprivation from spending hours on end in the newsroom or behind a microphone. Whether you are in the newsroom or in the field, don't let winter storms sneak up on you.

MAKE WINTER STORM COVERAGE PART OF YOUR NEWSROOM'S CULTURE—NOW

Begin preparing your staff for winter storm coverage weeks, months, or even years before one happens. Take every opportunity to explain in detail how covering moments of crisis—including winter storms—helps define your station as a news leader, how it's part of the very fabric of your news operation. Instill in your staff that it's critical that they provide winter weather forecasts, road condition reports, school closings, and other critical information—around the clock—when winter storms hit. Reinforce this notion often. Give staff members the chance to opt out, by finding another job, well in advance. That way neither they, nor you, will be surprised at the worst possible moment, when the snow begins to fall.

HAVE A PLAN

It's clichéd, perhaps, to suggest that you need a plan. But that makes it no less necessary. If you know in advance how many reporters you'll have in the field, who'll be on call for anchoring or newsroom duties, etc., the experience won't be quite so discombobulating. Make sure that your plan is comprehensive, that it anticipates every conceivable contingency. You can learn more about what to include in a plan by reading Elliott Wiser's chapter (see chapter 15). Most of his suggestions apply to covering severe winter weather. Make sure, too, that your plan has plenty of flexibility built in. Having a plan that makes no accommodation for changing conditions is as bad as having no plan at all.

Have a Snow Closing System Ready

If your station is at all like mine, collecting and reporting snow closing information will be one of the most resource-consuming aspects of covering harsh winter weather. It's also one of the most important to the audience and, therefore, one of the most crucial to establishing or maintaining your status as a news leader in your market.

My station is lucky. We have a sister television newsroom to pitch in with support staff when it's time to write or call schools and businesses to encourage them to participate. We have a strong information technology department to help us collect and then disseminate the list of closings on the air and on our Web site. If you aren't so blessed, start working now to develop relationships that will make this critical endeavor less stressful on snowy days. That includes recruiting people from every department in your building to help.

Get Your Vehicles Ready

Make sure your station vehicles are maintained properly, with adequate antifreeze and other fluids, and with appropriate, properly inflated tires. Invest in full-sized spare tires for each of your vehicles. Each should also be equipped with a snow shovel, a first aid kit, blankets, emergency provisions such as bottled water and non-perishable food (see the "emergency food" list in chapter 7), and even kitty litter (for enhanced traction). If some of your station vehicles are four-wheel drive, develop a schedule of who'll drive them to shuttle other staff from their homes. If you don't have any four-wheel drive vehicles, enlist employee volunteers who do have them to get your staff to work. If none of your employees has a four-wheel drive vehicle, hire someone now to transport your staff when the winter storm hits.

Get Your Station Ready

Have a few cots and some blankets stored somewhere in or near your newsroom so people who may not be able to leave for a few days have a place to try to get some rest. You should also identify a quiet office or conference room you can turn into a makeshift sleeping room. Or, if you're lucky enough to have a hotel or motel very close to your station, book rooms in advance so your staff has somewhere they can safely walk to get away from the newsroom for a few hours. It's also important to have bottled water and food available.

Get Yourself Ready

In my experience, it is extremely rare to be surprised by a winter storm. Forecasters usually suspect that it might happen at least a few days in advance. The night before the snow is supposed to start falling, eat a good meal. Relax a little. Try to distract yourself for a couple of hours with a book or a movie. Get plenty of sleep. It may be your last opportunity for quite some time. I realize all of this will be hard to do for aggressive news directors or on-air staff who work in a competitive environment. But, believe me, you'll be better off if you do it.

EXPECT THE UNEXPECTED

No matter how thorough your plan and your general state of preparedness, covering the winter storm is going to be stressful. Something you don't expect and can't anticipate *will* happen. Realize that going in. Whenever I acknowledge that a stressful situation is about to happen and I steel myself for it mentally, it somehow doesn't seem quite so stressful while it is actually occurring.

ENCOURAGE YOUR STAFF TO BE CAREFUL

Remind your staff, particularly those who will be in the field during the winter storm, to slow down. Reinforce this point so often that your staff accuses you of being a mother hen. They may tease you, but they'll remember your message. The adrenaline rush that generally helps them be such good broadcast journalists can be hazardous to their health in this context. No story is so important that they can't take just a moment to make sure they're being safe.

REWARD EVERYONE WHEN IT'S OVER

One of the greatest truisms of our business: that which is praised is repeated. One of the saddest truisms of our business: relatively few news directors praise their staffs. It's easy to single out employees who do extraordinary work. Taking a moment to do so after every winter storm, when it's deserved, will help ensure the next round of winter storm coverage goes just as well or perhaps even better.

I no longer find myself careening down an icy embankment, or cracking my leg falling down snow-covered stairs, or getting sunburned standing next to the flames of a burning house to avoid frostbite. In fact, very little surprises me any more when it comes to covering harsh winter weather. I suppose that's a manifestation of my longevity in this business, not my expertise. But one thing that does surprise me, in a pleasant way, every single time, is the drive, determination, and hard work of my staff. You will be pleasantly surprised, too, if you take the time—now—to make sure your newsroom is prepared.

Dan Shelley is news director and assistant program director for Newsradio 620 WTMJ in Milwaukee. He has been a broadcast journalist since 1980 and a news director since 1983.

LESSONS FROM THE WARS

In the beginning . . .

On Sunday evening, March 13, 1938, CBS Radio announcer Robert Trout brought listeners firsthand reports of Hitler's invasion of Austria. This marked the first time a group of correspondents in the field was linked by a central anchor. Thus was born the "CBS World News Roundup," and with it the era of broadcast war correspondents.

For the next seven years millions of Americans "watched" the radio as Edward R. Murrow and his legendary team of correspondents brought the drama and tragedy of World War II into American homes.

By the time journalists saw war again, in Korea, television was part of the equation. In the 1960s, Vietnam became known as the "living room war," with television coverage offered in graphic detail on the evening news each night during the dinner hour. And, by the time the United States carried its "war on terrorism" to Afghanistan and Iraq, technology made 24/7 coverage by satellite a reality and we witnessed the creation of the "embedded" journalist.

Michael Freedman

CHAPTER

17

War Stories

by Michael Freedman, Vice President and Professorial Lecturer in Journalism, The George Washington University, Washington, D.C.

"I had a full sense of the danger of war and had some concern as to whether I would return or not."

Walter Cronkite

While advancements in technology have dramatically altered the speed and methodology of the delivery of news, the personal experiences of war correspondents have remained remarkably similar from World War II to Iraq. This is evidenced by the recollections and views of veteran journalists Walter Cronkite, Richard C. Hottelet, George Herman, Liz Trotta, Marvin Kalb, and Cami McCormick. All say you can never adequately prepare for covering war. All have lingering memories. And all did their best to put their emotions on hold in order to cover the story.

Cami McCormick spent nearly three months in the Persian Gulf during 2003 reporting for CBS Radio Network News, first from Bahrain, then from aboard an aircraft carrier, then from Doha (Qatar), Kuwait, Jordan, and various points in Iraq, and finally from Baghdad. Prior to her departure for the Gulf, she participated in the official military training for journalists and then took a private course taught by British commandos in the mountains of Virginia. Ultimately, however, she felt it was her previous experience in the field—including being shot at in the Middle East—that came closest to preparing her for the realities of the war.

Once in the situation, says McCormick, you put your reporting before your fears. "Usually, I'm not frightened until afterward," she says. "I use everything I can to keep going and it's not until afterward that I feel the stresses of what I'd been through."

And how did people deal with the stresses during the war in Iraq? Exercise? Alcohol? Talking it through? "Probably all of the above," says McCormick. "And, again, it depended on where I was. In Doha, where a lot of journalists were gathered, and they were very frustrated because it was a place where you weren't getting the information that we thought we were going to get, there was a lot of drinking, a lot of eating, a lot of talking. But this is probably the first situation I've been in where there was also a lot of exercising. I saw a lot of journalists organizing squash games—anything they could to get some sort of physical activity to deal with, at least in Doha, just the monotony of it, every day."

Former CBS newsman George Herman says he didn't need to exercise in Korea. "I was carrying a sixty-five-pound tape recorder up a hill and down a hill and everywhere I went. And so there was no problem of needing exercise, for me." And he had no trouble sleeping, either. "You have to remember," he says, "that I started out in this business working the overnight shift. And after you've worked the overnight shift a year or two, you can sleep anywhere, anytime!"

BUT MEMORIES LINGER

"It seems to me I can remember almost every minute I was under fire in one place or another; even the conversations that went on come back to me," says Walter Cronkite. The dean of American television journalists attributes this to the level of professional concentration necessary to cover the story. "You know what it is? That's the old business of being able to concentrate so completely on something that it records in this amazing computer that is our minds. It's like athletes who can tell you every stroke of a golf game that they won; a tennis player who, with all of those shots that he or she is making, can go back to a championship match ten years earlier and tell you what the score was in the third game and how they served a certain serve and how they met the challenge. The mind is really working overtime in situations like that. There's no opportunity for a little laziness that would let those memories go unrecorded," says Cronkite. And that goes for memories both good and bad.

OLD IRON PANTS

Cronkite earned his journalistic stripes at a wire service, covering World War II for the United Press. He witnessed the battle of the North Atlantic, landed with the invading Allied Troops in North Africa, and took part in the Normandy beachhead assaults on D-Day in 1944. He was among the first newsmen to participate in the B-17 bomber raids over Germany, dropped with the 101st Airborne Division in Holland, and was with the U.S. Third Army during the Battle of the Bulge. His unflappability under pressure earned him the affectionate nickname "Old Iron Pants." And all this took place *before* Cronkite joined CBS News in 1950, where he set the television standard for poise, trust, and respect.

The battle of the Atlantic was at its peak when Cronkite became one of the first correspondents to receive credentials to cover the Navy and sailed off on his first convoy. He had no delusions of invincibility. "I had a full sense of the danger of war and had some concern as to whether I would return or not," says Cronkite. "But I think each individual reacts to that kind of danger differently. At that time in my life, I just felt it was part of the job. I was sort of proud of undergoing the experience. I felt this was what I should be doing."

AN OCCUPATIONAL HAZARD

Like Walter Cronkite, Richard C. Hottelet was a United Press wire service correspondent, before joining Edward R. Murrow's legendary team at CBS Radio. While with the UP in Berlin in March 1941, he became the only American journalist arrested by the Nazis. Hottelet spent four months in solitary confinement, held on trumped-up charges of suspicion of espionage. He was ultimately released in a prisoner exchange. He was twenty-three years old at the time.

Hottelet's take on his four months of captivity at the hands of the Gestapo? He terms it simply "an occupational hazard." Hottelet says that, in a classic journalistic postmortem, when he returned to New York after being released, one of his bosses at United Press first tried to calculate his back pay without the meal money he was owed, citing the fact that he had been fed in prison!

Three years after his arrest by the Gestapo, at age twenty-six, Hottelet was hired by Murrow. Along with his fellow "Murrow Boys," he helped write the book on broadcast war coverage. Among his many battle assignments, Hottelet flew in a bomber that attacked Utah Beach on D-Day, and broadcast the first eyewitness account of the seaborne invasion of the Normandy coast.

Of his work in World War II, Hottelet says he wasn't fearful but points out it was "not a question of bravery" but rather one of environment. "If you are in the field with an army—or in a plane with a crew—you are just one of several who are sharing what may be a scary experience, and the sharing means that your own fears are, sort of, subsumed in a community of anxiety, and they don't bother you."

Hottelet and Cronkite, like many members of what we now refer to as "the greatest generation," refuse to take any individual credit for their coverage in battle. "What we did was simply translate the normal standards of reporting in civil life to a military situation," Hottelet says. And of the fears of being in battle zones? "You did your job through the anxiety."

George Herman admits feelings of doubt about his own ability to "cut the mustard" in Korea, because his experience to that point had been as a news writer for CBS. He had the added challenge of following in the footsteps of the legendary Murrow Boys of World War II.

The young writer, who would later serve as a CBS News White House correspondent and moderator of *Face the Nation,* covered the Inchon invasion in the fall of 1950 and remained in Korea and Tokyo, site of the UN Headquarters, until the armistice was signed in 1953.

Herman quickly achieved a comfort level with his new role but admits to occasional fears for his personal safety. "I once realized that I had accidentally strayed between the battle lines," he recalls. "I never knew a Jeep could turn around so quickly and come back to our own battle lines." He also faced another dilemma. "I was concerned about some of the things I was seeing our troops do," says Herman. "It had never occurred to me that American troops could commit—well, let's be fair and call them 'atrocities'—as well as foreign troops could. And I didn't quite know what to do with that. How do you report, in the middle of a war, which we weren't necessarily winning a lot of the time, that you had seen American troops do some dreadful things that made you a little sick to your stomach? That was a difficulty for me."

How did he deal with it? "On the air, I ignored it," says Herman. "If somebody reported it, if there were an official document on it, I'd cover that. But just things that I saw that were unpleasant—theft or rape or what have you—I tended to ignore it. I did not report it."

Herman says he valued the close relationships developed with other journalists in Korea and he talked things through with them "when something happened that fascinated or disturbed me. And frequently," he adds, "I would find, to my encouragement, that they were treating it the same way I was—ignoring it or toning it down or giving it big play."

Such was *not* the case for Liz Trotta, television's first female war correspondent. Still in her mid-twenties, Trotta had already served as a local newspaper and television reporter when she received her first war coverage assignment, in 1968, from NBC News. Trotta says she didn't just volunteer—she actively campaigned—to be sent to Vietnam. There, she says, she found the troops, but not her journalistic colleagues, grateful for her presence. "The biggest problems I had came from members of the press," says Trotta. "The only people who gave me trouble were my fellow correspondents who could not compete with a woman. They just couldn't do it." The exception, she says, was then–Associated Press correspondent Peter Arnett, whom she credits as being "the one person who was quite fair to me—of all of them. Peter had always treated me very well. But that was an exception."

Having faced everything from full-scale conventional large engagements, sniper fire, booby traps, and the horrifying sight of injured children, to hostile colleagues and the elements, Trotta says her time in Vietnam was rife with very real concerns. "I was scared out of my wits all the time," she says. Still, the tough, independent-minded Trotta, who went on to cover many other conflicts, including the India-Pakistan war, the violence in Northern Ireland, and the Yom Kippur War in the Middle East, had no interest in making friends by "becoming one of the boys." "No," she says, "I lived like a nun—but then I had been trained by nuns." How did she cope? "I learned how to fade, and I tried to be on my best behavior at all times. I prayed a lot," she adds, "and that helped." But what made the biggest difference was the journalism. "The work saves you," says Trotta, now New York Bureau Chief for the *Washington Times.* "The tyranny of the deadline is a great discipliner."

STRESS AND REPORTING GO HAND IN HAND.

"Stress is part of good journalism," says former CBS News chief diplomatic correspondent Marvin Kalb. "Stress comes with the very act of total engagement in a story," he adds. "Are you going to be able to get the story? Are you going to be able to get it right? Are you going to be able to get it quick enough for it to be meaningful? Are you going to be able to report it with eloquence? These are questions that are associated with the business but they are also questions associated with layers of stress that cannot be described."

Today, adding to the stress of getting the story and getting it right is the challenge of the twenty-four-hour news cycle, says Cami McCormick. "It does add a lot of pressure to you as a journalist, and it perhaps makes you take chances you wouldn't take otherwise, just because there is simply that pressure to constantly know what's going on and to constantly be reporting because there's a kind of twenty-four-hour news monster to feed."

THE HORRORS OF WAR

Fifty years after Korea, George Herman vividly recalls that "when the Chinese came in and drove down to recapture Seoul after we had cleared it of North Koreans, the American Army moved out of Seoul, bag and baggage, and blew the bridges, leaving the Koreans behind, still in the city. I thought it was dreadful. I watched the Koreans wade across the frozen river—sometimes on the ice, sometimes in the slush—to the railroad station across the river. And since the trains were full, they tied themselves to the outside of the trains with straw ropes and would be taken off at the next station—frozen solid. And I saw grandmothers trying to get rid of the grandchildren, because there was no way they could take them, and the American GIs had to try and gather up the babies—it was a shameful, hurtful, embarrassing scene, and I was temporarily embarrassed to be an American. That sticks in my mind," says Herman.

McCormick says a memory that will remain with her is "what we saw going into Baghdad . . . the damage that was done, the cars still smoking on the highway leading into Baghdad. Someone turned around and said 'There were bodies in there just yesterday.' That kind of situation, the chaos, the stench on the streets, the total chaos in Baghdad when we arrived, and the smells and the people. That, I'll never forget as long as I live."

Hottelet says his memories are vivid enough but adds that reporters need to function like surgeons in these situations. "You can agonize over every incision if you want to and wear yourself out, or you can just add it to your experience," says Hottelet. "And I think in my case, and in most others, it was not internalized in any disturbing way at all, but it just enriched your background."

It was concern for others that left an indelible mark on Marvin Kalb, who recalls an incident that occurred in 1964 when he was sent to Saigon for *Face the Nation.* Marvin's brother Bernard Kalb, also a CBS News correspondent, was already in Vietnam at the time. Marvin, Bernard, and a crew of producers and technical staff were driving in a convoy of cars in the countryside south of Saigon when shooting erupted on the right side of the road. As the firefight was ending, first the crew and then Bernard Kalb got out of their vehicles to cover the action.

"I remember screaming at him to come back," says Kalb. "He, of course, went ahead and did his job." The story got covered and the incident ended without harm to the CBS News crew. But it lingers in the memory of Kalb—not because of what might have happened to him, but out of the deep concern he felt for his mother. "What was in my mind," says Kalb, "was 'What would New York tell Bella Kalb if both of her boys were caught in a firefight and never made it out?'"

FROM V-MAIL TO E-MAIL ...

Walter Cronkite was also more worried about his family back home than himself. "That was the worst part," he says. "They suffered concern far greater than I did. I knew what the nature of the concern was. All they knew was that I was somewhere in a war zone."

For Cronkite during World War II, communication with loved ones occurred though V-mail, a device that reduced a full letter to the size of a postcard. These were transported back to the United States from the front lines. "My feeling always was to try to bolster their spirits and morale," says Cronkite, "rather than their bolstering mine, by the return of V-mail letters."

Herman bolstered the spirits of many families of soldiers in Korea by having "CBS NEWS" printed across the front of his Jeep. After interviewing servicemen he would get names and phone numbers of relatives and pass the information on through his cue circuit so families could be notified that their loved ones were okay. "I purposely went out to make the CBS Jeep welcome everywhere we went," says Herman. He adds that these favors also cultivated a general feeling of welcome for CBS News.

Fast forward to 2003 in Iraq, and journalists had a little easier time staying in touch with friends and family. "Thank goodness for e-mail," says McCormick. "It's changed a lot of things."

But some things don't change. War is hell. People die. And journalists are affected by it.

Even Cronkite, with more battles under his belt and more brushes with death than most of today's journalists *combined,* still wrestles with a single decision he made more than half a century ago. "I never ducked an assignment," says Cronkite. But he says he once turned away an *opportunity* during the Battle of the Bulge, and his rationale for doing so has concerned him ever since.

The 101st Airborne Division was surrounded in Bastogne, encircled by the Germans. Cronkite was with an armored unit that was leading the spearhead trying to get through the German lines. General Maxwell Taylor, Commander of the 101st Airborne Division, had been in the U.S. when his troops were surrounded. He flew back immediately to try to rejoin them inside. After a couple of scouts were able to get in and out of the city—under heavy fire—General Taylor decided to make his move. "Okay, Cronkite, I'm going to Bastogne," he told the young war correspondent. "Do you want to go?" Cronkite says he started for the Jeep but realized that if he went *in* there would be no way to get the story *out.* He decided it was smarter to cover the story from his current vantage point than go into Bastogne and not have any communication at all.

Today, Walter Cronkite still wrestles with that call. "I don't know whether it occurred to me at that point that I'd get killed on that trip, it was foolhardy, or if I came up with the rationale later that if I went in there with [General Taylor] I had no way to get any story out of Bastogne," says Cronkite. "To this day, I'm not sure which was the dominating influence."

McCormick and Trotta say, for them, battles infiltrated even sleep. "When I went to the West Bank and covered the fighting there," says McCormick, "I had bad dreams for about a month afterward and I had to deal with those issues. But at the time you just do what you can to stay alive and cover the story." Trotta says while in Vietnam she had a lot of bad dreams but they were not lasting. Among the memories of the many places where Trotta covered stories, Vietnam does stand out, however, "because it was such a huge story and such a vivid story. And you see it in Technicolor," she says, "even though the film was black and white."

McCormick tells aspiring reporters to "prepare for anything and go and experience as much as you can now, because no one can really teach you a lot of these things. You've got to get out there and do it."

Hottelet encourages journalists not to become part of the story. "Don't dramatize yourself or the situation," says the legendary Murrow team member. "Just keep your feet on the ground and be as objective as you can," Hottelet adds. "Of course, put your own emphasis to it, but just play it straight."

Trotta reminds journalists that "being in combat is a high." And she warns "that high can dissolve into some miserable stuff if you're under siege for a long time."

Herman says the first order of business is to stay alive. "Watch your head and duck," says the former CBS News correspondent. "Be ready to endure the sight of dreadful things—blood and guts and death. Be ready to be sad for sad things, and be ready to translate it into clear, moving prose." And George Herman also makes a plea to take care of your colleagues. "Protect your cameraman and where you can; make sure he or she gets some credit for guts."

Cronkite believes each individual responds differently to war. "I have colleagues who weren't particularly anxious to get up at the front at all, who went where the battle was, where the shooting was, rather reluctantly and sparingly," he says. "Others were far braver than I and were out there every day." But, Cronkite adds, in his case he always looked forward to the next good story, "and the next good story was something in combat; and therefore, I was always volunteering, and not always without trepidation. In fact, I'd say, always *with* trepidation."

Michael Freedman is a vice president and professorial lecturer in journalism at The George Washington University in Washington, D.C. He is the former general manager of CBS Radio Network News in New York and managing editor for the broadcast division of United Press International in Washington. Over the course of his thirty-year career as a broadcast journalist, network executive, university administrator, and teacher, Mike and his teams have received more than eighty-five awards for broadcast excellence, including fourteen Edward R. Murrow honors from the Radio-Television News Directors Association.

"The ultimate measure of a man is not where he stands in moments of comfort and convenience, but where he stands at times of challenge and controversy."

Martin Luther King, Jr.

"I am not afraid of storms,
for I am learning how to sail my ship."

Louisa May Alcott

CRISIS SURVIVAL KIT

"Hope for the best and let that dominate most of your emotions, but at the same time let one small piece of you prepare for the worst."

Reinhold Niebuhr

"Being challenged in life is inevitable, being defeated is optional."

Roger Crawford

CRISIS SURVIVAL KIT 1

Coping with Stalkers: A Safety Checklist

Adapted from *Broadcast Voice Handbook* by Ann S. Utterback

THE HIDDEN CRISIS

One of the best-kept secrets in every newsroom is that stalkers are a serious problem. Everyone knows it, but no one talks about it. Stalkers create a hidden crisis for people, because the stress is tremendous but it is often suffered alone.

A stalker is a person who intentionally and repeatedly harasses another person, provoking fear in that person. The unwanted contact can be in the form of phone calls, letters, gifts, or personal encounters. These contacts are normally not hostile in the beginning, but what begins as a nuisance can quickly escalate into a threat.

On April 23, 2003, a segment of the TV newsmagazine *48 Hours Investigates* (produced by Martin Zied) focused exclusively on broadcasters who had been stalked. In the report, one of the country's foremost forensic psychiatrists, Dr. Park Dietz, said that the question is not *if* a celebrity has a stalker; it's *how many* he or she has. According to Dietz, "Nobody knows right now how many news anchors get stalked. The likelihood is that, for good-looking females on the news nightly, it's 100 percent."

The Justice Department estimates that more than one million women are pursued and harassed by stalkers every year. In *Currents in Affective Illness* (1992), Dr. Dietz writes, "The nicer one appears to be, and the more approachable, the more one will attract the serious and persistent and deluded subjects." He believes that the perceived approachability and the high visibility of broadcasters make them prime targets for stalkers. His company sees a couple dozen cases of stalking involving broadcasters every year.

I have had many clients tell me about being stalked. There is nothing more stressful than thinking your life is in danger. The mental anguish this creates can be paralyzing. When you are the victim of a stalker, it's like living in a house with no curtains. The stalker can watch you simply by turning on the television.

In her book *Surviving a Stalker,* Linden Gross says, "Stalking is like a long rape. The stalker's objective is to force you into surrender." Broadcasters have left the business because of a stalker, and some have even been killed.

Here is a safety list to help you safeguard yourself and cope with any stalker situation that might arise.

SAFETY CHECKLIST

If You Begin to Feel Uncomfortable

- ❑ If you receive two or three suspicious calls, letters, or gifts from the same person, or if the same person approaches you on several different occasions, contact the police. No threats need to have been made.
- ❑ Do not talk to the person yourself to try to clarify things or make them understand your feelings.
- ❑ Do not return gifts or letters. Save these and any telephone messages for the police.
- ❑ Document any contact with the person.
- ❑ Do not eat any food items sent from a stranger.
- ❑ Do not open any unexpected packages or letters. Take them to the police.

When You Move to a New City (or If a Stalker Becomes a Problem)

- ❑ An on-air name that is different from your real name makes security and safety much easier. Many broadcasters use on-air names. If you choose to use your own name on the air, consider the suggestions that follow.
- ❑ Request an unlisted and unpublished telephone number. Get caller ID on your phone. Request a complete blocking order (or per-line blocking) for your own phone line so your number will not be revealed on others' caller ID.
- ❑ Use a private mailbox service to receive all personal mail. You can list your box number as an apartment number. Make certain that the mailbox service will not release your address to anyone other than law enforcement personnel. Send a change of address to friends and businesses. File for confidential voter status.
- ❑ Write to the credit reporting agencies and give them your new address. Ask that they remove your old address from any past credit history. You might also want to ask the three main credit bureaus—Equifax *(www.equifax.com),* Experian *(www.experian.com),* and Trans

Union *(www.transunion.com)*— to flag your account to alert you to any fraudulent access.

- ❑ Avoid putting your home address on personal checks or business cards. Use the station's address. Pay with a credit card in your community to avoid showing your ID.
- ❑ Never give out your own phone number or address over the phone or in a store. Use your number and address at the station.
- ❑ If renting an apartment, place the rental agreement in another person's name. If you must use your name, ask the landlord not to give your name to the city directories. Do not rent an apartment on the ground floor or one that has easy access through windows or balconies.
- ❑ Have a friend or reputable locksmith install deadbolt locks on all doors. Secure all sliding glass doors and windows with locks. Keep all doors and windows locked. If possible install extra outside lighting.
- ❑ Demand identification from all repairmen and salesmen prior to entry. Even ask the police for identification prior to entry. Ask neighbors not to give out any information about you to anyone.
- ❑ Do not put personal information on the station's Web page. Imagine that what goes on the Web page is the same as what you'd share in an Internet chat room with someone who seems too interested in your personal life. Keep your comments very general.
- ❑ Ask that your name and address be left off any address list made up for station employees. Alert the station employees (especially the receptionist, assignment desk, and camera people) not to give out personal information about you such as your birthday, address, or marital status. If you are single, you might consider wearing a wedding ring on the air to give the impression that you are married.

When You Are Commuting to Work

- ❑ Be alert! Many attacks happen in cars going to or coming from home or work. Watch for vehicles that might be following you. Always keep at least half of a tank of gasoline in your car.
- ❑ Have an exact time you are to arrive at work and check in with the same person every day when you arrive. This is especially important if you work an early morning shift. If you are driving in the early morning hours, use a cell phone or radio to talk to someone at the station while you drive in.
- ❑ Park in a different place every day both at home and at work and vary the route you drive to work. If you will be leaving after dark, park in a well-lit area. Get someone from security or master control to walk you to your car.

- ❑ Avoid driving an easily recognizable car and do not use a vanity license plate. Keep your car doors locked at all times both when unattended and when you are driving. Look into the front and rear of your car before entering the vehicle.
- ❑ Notice where police stations, hospitals, fire stations and other public places are on your driving route. If you are being followed, drive to a well-populated area like a hospital or fire station and sound your horn to attract attention. Remain in your locked car. Use your cell phone to call for help.

Covering a Terrorist Incident—Advice from Terrorism Experts

by NewsLab

This article is reprinted with permission of the executive director of NewsLab, Deborah Potter. She is a former network correspondent. You can find out more about NewsLab at www.newslab.org.

There's no way to know all there is to know about every possible terrorist threat. But every journalist should know something about what one expert calls B-NICE: biological, nuclear, incendiary, chemical, and explosive hazards. Here are some tips from experts who spoke in June 2003 at the "Media and the First Response" conference in Washington, hosted by the National Academies.

BE PREPARED

Know the potential targets in your area. Government buildings, military installations, storage facilities, plants, public buildings, places where large groups gather.

BE ALERT

Notice the out-of-the-ordinary, and know what it could mean.

- Something obvious like an explosion is probably a conventional attack. If it's a dirty bomb (an RDD, or radiological dispersion device), says George Whitesides of Harvard, you are more likely to be run over by somebody trying to get away than to be hurt by the radiation. "A dirty bomb sounds terrible, but the risk is moderate."

- Are there people down? Likely a chemical attack. Stay away. Stay upwind.
- Unexpected incidents of disease in hospitals? It could be biological, but be very careful with reporting until something is confirmed.

BE CAREFUL

If you hear gas escaping, you are too close. If you smell an unusual odor (like newly mowed hay, corn, garlic, camphor, or bitter almonds), you may have been exposed and may be contaminated. Avoid further risk by retreating, says Warren Campbell of the Maryland Fire and Rescue Institute. Do not eat, drink, smoke, or chew gum anywhere near the scene. You could ingest the substance.

Escape hoods and masks are great devices for one thing only: escape. These are not for reporters to put on in a clean area and go into a dirty area. That requires training, fit testing, and different suits for different classes of agents (from A, the highest protection, to D, the lowest).

CHANGE YOUR REFLEXES

Think of the situation as an unfolding event, not just as a story to cover. You in the news media may be a primary target. Terrorists want to take out the first responders, including journalists.

TAKE ACTION

If you have droplets on you, blot or scrape them off immediately. Don't scrape hard enough to break skin. If you've been exposed to a blister agent, seek treatment. Don't damage the blisters. If you're dealing with biological agents, don't share food or drink. Keep trashcan lids on tight and remove standing water, or varmints and insects can spread the damage. If dealing with radiological contamination, safety can depend on time, distance, and/or shielding. Avoid inhaling dust or smoke. Shower and wash hair thoroughly.

BE INFORMED

Exposure and contamination are different. If you walk through a gas there is no need to decontaminate. Only a liquid or solid will stick. The effects of an attack depend on the substance involved, the route of exposure, and the level of exposure. The poison is in the dose. If the route is skin absorption, the weather matters. On a hot day, the rate of absorption is faster. Effects are cumulative. Just because you aren't affected by one exposure doesn't mean you won't be affected.

KNOW THE DRILL

First responders will typically set up three zones:

1. Hot: in the immediate area of contamination
2. Warm: not contaminated yet, but it likely will be
3. Cold: safe

They will deny entry to both hot and warm zones. A command post will be set up in the cold zone (no suits or masks required). They will establish an access corridor to send people in and a decontamination corridor to get people out.

The incident response team will take these steps:

1. Look for a second device
2. Remove casualties through triage and send them for medical care
3. Detect and identify what they're dealing with, taking samples and testing (not a fast process)

The incident commander can decide to send reporters into the zone and will provide them with appropriate gear. Problem: it's likely to be a one-way trip for the equipment. It's just not clear how you would decontaminate a camera.

KNOW WHAT TO ASK

Ask officials for a timeline, and ask for it now. Make them tell you when they'll provide what kind of information in a crisis.

Phase your questions to improve your chances of getting answers.

- Public health: what is the hazard and how can people protect themselves
- Environmental/economic: what is the damage and how can it be cleaned up
- Intelligence/national response: what does it mean and what will we do about it

Ask better questions to get better answers, says Margaret Hamburg, VP for biological programs at the Nuclear Threat Initiative. Example: don't ask when a location will be cleaned up; ask when it will be clean enough to not endanger public health.

BE USEFUL

People are going to make their own decisions. "It's a Washington fallacy to think people will do as they are told," says Jay Davis of ANSER. Give people useful information that will help them decide what to do. For example, he suggests this simple rule of thumb:

- If the glass is broken in your building, leave
- If the glass is intact, stay there until told otherwise

AVOID THE BLAME GAME

When information turns out to be wrong, it doesn't mean that officials are either scoundrels or incompetent, says Davis. You can draw different conclusions when you have more data.

CHEMICAL ATTACK

A chemical attack is silent. It doesn't start with a bang. Symptoms become apparent, leading to fear, confusion, and incapacitation.

Gases are heavier than air so the plume moves along the ground. Avoid low-lying areas.

Protective gear is bulky and may not help. You don't know what mask to put on unless you know exactly what you are dealing with.

If an attack is indoors, it may not be clear whether ventilation will help or hurt. Do you turn the system off or on?

BIOLOGICAL ATTACK

Don't be misled by our experience with anthrax. This was much more like dealing with a hazmat situation than with a biological attack. It came with a note, at a specific time and place.

A biological attack is not a "lights-and-sirens" event. There would be no announcement, no signal, no area to cordon off and clean up. The site of release might never be identified.

First responders are not police and fire but public health workers, emergency rooms. Ground zero would be clinics and labs.

The experts:

- **Warren Campbell, Maryland Fire and Rescue Institute**
- **Jay Davis, director, ANSER Institute for Homeland Security**
- **Baruch Fischhoff, department of engineering, Carnegie Mellon University**
- **Margaret Hamburg, VP for biological programs, Nuclear Threat Initiative**
- **Alice Gast, professor of chemical engineering, MIT**
- **George Whitesides, professor of chemistry, Harvard**

CRISIS SURVIVAL KIT 3

Go-Bags Essentials

In addition to what you normally carry with you to work or in the field—such as water, your cell phone, IFB, makeup, and PDA—it's helpful to carry this *emergency fanny pack* with you at all times in case of a crisis event. A complete *go-bag* can be kept in your car or at the newsroom to provide what you will need for survival in an extended crisis. Grab it along with the fanny pack any time you're sent on an assignment that might last overnight.

EMERGENCY FANNY PACK

Carry this on all assignments.

- Swiss Army knife or Leatherman tool that includes a can opener and scissors
- $10 roll of quarters and $50 to $100 in $1 and $10 bills (ATMs will not work if power is out. You may need quarters for pay phones.)
- Small battery-operated FM/AM/TV radio (store batteries outside radio)
- Small flashlight (store batteries outside flashlight)
- Small packet of facial tissues
- Small antibacterial hand cleaner
- Protein bar
- Disposable lightstick (will provide light for 8 to 10 hrs.)
- Pen and Sharpie marker
- Pain reliever and extras of any prescription drugs
- "In Case of Emergency" card with contact numbers
- Phone number and e-mail address of out-of-town contact you and your loved ones have agreed to contact should a disaster strike and the local phones are out

GO-BAG

Large backpack or athletic bag. ***Carry this along with your fanny pack for assignments that may last overnight.***

- Fully-charged cell phone batteries
- Extra pager batteries
- Maps and compass
- *Food items to be refreshed after six months of storage:*

  Two 50 oz. bottles of water
  Jar of peanut butter
  Canned meat and tuna, and dried fruit
  Protein or granola bars
- Plastic utensils
- Moistened towlettes
- Several small packs of facial tissue
- Leather and latex gloves
- First aid kit
- Sunscreen
- Duct tape
- Large, folded plastic trash bag
- Large flashlight (store batteries outside flashlight)
- Paper copies of important phone numbers both personal and professional; 800 numbers for hotels, car, and airlines; copy of passport
- Pair of comfortable, sturdy shoes
- Rain jacket or poncho
- Reflective emergency runner's blanket
- Baseball cap without a logo (except a station logo)
- Small toothbrush and toothpaste
- Small shampoo
- Deodorant
- Brush/comb
- Extra pair of glasses if needed
- Change of underwear and clothes if space permits

Suggestions for this list came from David Steck (see his complete list in *Broadcast Voice Handbook*), the American Red Cross literature, and the author's own experience.

CRISIS SURVIVAL KIT 4

Emergency Food for Busy Broadcasters

CARBOHYDRATES

These can be stored in your desk or bag in a plastic or metal container. Check your storage items once a week to restock and dispose of old or spoiled food.

- Low-fat granola bars
- Small boxes of raisins or dried fruit
- Canned or dehydrated soups
- Pretzels
- Vanilla wafers
- Baked or low-fat chips
- Instant oatmeal
- Breadsticks
- Individual boxes of cereal
- Low-fat microwave popcorn
- Graham crackers
- Rice cakes
- Fig bars
- Oyster crackers
- Juice boxes
- Small cans of fruit in light syrup
- Snack-packs of applesauce
- Vegetable juice in small cans
- Whole grain breads or crackers
- Pita bread

- Whole wheat bagels
- Animal crackers
- Vegetable sticks (store in fridge)
- Fruit (Oranges, pears, bananas, grapes, and apples are easy to store and travel well. You might want to keep these in a bowl on your desk as a reminder to eat!)
- Baking potatoes (These make a quick meal when microwaved and topped with some protein.)

PROTEIN

Most of these items will need to be refrigerated or kept in a lunch bag with a cold pack. Check food in the fridge once a week and discard spoiled items.

- Nonfat yogurt
- Low-fat cream cheese
- Low-fat milk
- Low-fat cottage cheese
- Individual cheese slices or sticks
- Peanut butter (limit to two tablespoons a day)
- Slices of turkey, ham, or chicken
- Hard-boiled eggs
- Small cans of tuna in water or salmon or chicken (with pop-top opener)
- Protein bars (look for bars with less than 15 grams of sugar and at least 15 grams of protein—and some carbs, of course)
- Nuts (High in good fats; stick to a handful a day. Lowest in fat are almonds, pistachios, walnuts, and hazelnuts.)

To Prepare a Healthy Meal or Snack . . .

Think of this ratio when eating:

1/3 protein

with

2/3 carbohydrates

CRISIS SURVIVAL KIT 5

Caffeine Sources

Note: A mug holds around 14 oz., and many paper cups hold 20 oz. or more, so calculate how much caffeine you are actually getting based on the size.

Product	Average caffeine (mg)
Coffee	
Brewed, drip method (5 oz. cup)	115
Instant (5 oz. cup)	65
Decaffeinated, brewed (5 oz. cup)	3
Decaffeinated, instant (5 oz. cup)	2
Espresso (1 oz.)	40
Dannon Coffee Yogurt (1 cup)	45
Häagen-Dazs Coffee Ice Cream (1 cup)	58
Tea	
Brewed, major U.S. brands (5 oz. cup)	40
Instant (5 oz. cup)	30
Iced tea (12 oz. glass)	70
Lipton Bottled Iced Tea (16 oz.)	40
Green tea (8 oz.)	30
Chocolate Products	
Cocoa beverage (5 oz. cup)	4
Chocolate milk beverage (8 oz.)	5
Milk chocolate (1 oz.)	6
Dark chocolate, semisweet (1 oz.)	20
Baker's chocolate (1 oz.)	26
Small candy bar with chocolate	25
Hershey's Kisses (9)	5
Chocolate-flavored syrup (1 oz.)	4

Sources: FDA, Food Additive Chemistry Evaluation Branch, based on evaluations of existing literature on caffeine levels; *Self* magazine, October 2004.

Product	Average caffeine (mg)
Soft Drinks (12 oz. serving)	
Mountain Dew	54.0
Mello Yello	52.8
Coca-Cola	45.6
Diet Coke	45.6
Dr. Pepper	39.6
Sugar-Free Dr. Pepper	39.6
Pepsi-Cola	38.4
Diet Pepsi	36.0
Canada Dry Diet Cola	1.2

Source: Institute of Food Technologists (IFT), April 1983, based on data from National Soft Drink Association, Washington, D.C. IFT also reports that there are at least 68 flavors and varieties of soft drinks produced by 12 leading bottlers that have no caffeine.

Product	Average caffeine (mg)
Weight-Control Aids	
Dex-A-Diet II	200
Dexatrim Extra Strength	200
Alertness Tablets	
Nodoz	100
Vivarin	200
Analgesic/Pain Relief	
Anacin; Maximum strength Anacin	32
Excedrin	65
Midol	32.4
Vanquish	33
Cold/Allergy Remedies	
Triaminicin tablets	30
Dristan Decongestant tablets; Dristan A-F Decongestant tablets	16.2

Source: FDA's National Center for Drugs and Biologics. FDA also notes that caffeine is an ingredient in more than 1,000 nonprescription drug products as well as numerous prescription drugs.

How Well Am I Coping?

It helps to monitor how well we're taking care of ourselves. Here is a chart you can follow daily with a place where you can keep notes on how you're handling the stress in your life. Feel free to make copies of these pages for repeated use.

Date____________________

	Poor				Great!
Relaxation	1	2	3	4	5
Exercise	1	2	3	4	5
Letting Myself Sleep	1	2	3	4	5
Identifying My Feelings	1	2	3	4	5
Eating Well	1	2	3	4	5
Focusing on Fun	1	2	3	4	5

What did I do today to relax?

What exercise did I get?

How long did I sleep? Any problems sleeping?

Did I find ways to identify and express my feelings? What feelings came up for me today? On a scale of 1 to 10, how intense were these feelings?

When did I eat today? What did I eat that was healthy/unhealthy?

Did I schedule some time today for fun? What did I do?

Suggested Readings

American Cancer Society, "Cancer: Basic Facts," "Cancer Facts and Figures," *www.AmericanCancerSociety.org.*

Anderson, Bob. *Stretching at Your Computer or Desk.* Bolinas, Calif.: Shelter Publications, 2000.

Bailey, Covert. *The New Fit or Fat.* Boston: Houghton Mifflin, 1991.

———. *Smart Exercise.* Boston: Houghton Mifflin, 1994.

Beck, Martha. *Finding Your Own North Star: Claiming the Life You Were Meant to Live.* New York: Three Rivers Press, 2001.

Benson, Herbert. *The Relaxation Response.* New York: Avon, 1975.

Benson, Herbert, and Eileen M. Stuart. *The Wellness Book: The Comprehensive Guide to Maintaining Health and Treating Stress-Related Illness.* New York: Simon & Schuster, 1992.

Bogert, Fred. "What Will My World Be Today." In *Dreamspeak.* Maxbox Music BMI compact disc. Available at *www.chromerecords.com.*

Brody, Jane. *Jane Brody's Cold and Flu Fighter.* New York: W. W. Norton, 1995.

———. *Jane Brody's Nutrition Book.* New York: W. W. Norton, 1981.

Cameron, Julia. *The Artist's Way: A Spiritual Path to Higher Creativity.* New York: G. P. Putnam's Sons, 1992.

Carlson, Richard, and Joseph Bailey. *Slowing Down to the Speed of Life: How to Create a More Peaceful, Simpler Life from the Inside Out.* New York: Harper Collins, 1997.

Cherniske, Stephen. *Caffeine Blues: Wake Up to the Hidden Dangers of America's #1 Drug.* New York: Warner Books, 1998.

Chödrön, Pema. *The Places That Scare You: A Guide to Fearlessness in Difficult Times.* Boston: Shambhala Classics, 2001.

Davis, Martha, Elizabeth Robbins Eshelman, and Matthew McKay. *The Relaxation and Stress Reduction Workbook.* 5th ed. Oakland, Calif.: New Harbinger Publications, 2000.

Domar, Alice D., and Henry Dreher. *Healing Mind, Healthy Woman: Using the Mind-Body Connection to Manage Stress and Take Control of Your Life.* New York: Henry Hold, 1996.

Frankl, Viktor E. *Man's Search for Meaning.* New York: Simon & Schuster, 1959.

Gross, Linden. *Surviving a Stalker: Everything You Need to Know to Keep Yourself Safe.* New York: Marlowe, 2000.

Hanh, Thich Nhat. *Peace Is Every Step: The Path of Mindfulness in Everyday Life.* New York: Bantam Books, 1991.

Harris, Rachel. *20-Minute Retreats: Revive Your Spirits in Just Minutes a Day with Simple Self-Led Exercises.* New York: Henry Holt, 2000.

Inlander, Charles B., and Cynthia K. Moran. *77 Ways to Beat Colds and Flu.* New York: Walker, 1994.

Jacobson, Michael F., and Jayne Hurley. *Restaurant Confidential.* New York: Workman Publishing, 2002.

Jampolsky, Gerald, G. *Love Is Letting Go of Fear.* Berkeley, Calif.: Celestial Arts, 1979.

Kabat-Zinn, Jon. *Full Catastrophe Living: Using the Wisdom of Your Body and Mind to Face Stress, Pain, and Illness.* New York: Dell Publishing, 1990.

———. *Wherever You Go, There You Are: Mindfulness Meditation in Everyday Life.* New York: Hyperion, 1994.

Kundtz, David. *Stopping: How to Be Still When You Have to Keep Going.* Berkeley, Calif.: Conari Press, 1998.

Northrup, Christiane. *Women's Bodies, Women's Wisdom.* Rev. ed. New York: Bantam Books, 1998.

Ornish, Dean. *Dr. Dean Ornish's Program for Reversing Heart Disease.* New York: Ballantine Books, 1990.

Peeke, Pamela. *Fight Fat After Forty.* New York: Penguin Books, 2000.

Richardson, Cheryl. *Take Time for Your Life: A Personal Coach's Seven-Step Program for Creating the Life You Want.* New York: Broadway Books, 1999.

Sapolsky, Robert M. *Why Zebras Don't Get Ulcers: The Acclaimed Guide to Stress, Stress-Related Diseases, and Coping.* 3rd ed. New York: W. H. Freeman, 2004.

Schieffer, Bob. *This Just In: What I Couldn't Tell You on TV.* New York: G. P. Putnam's Sons, 2003.

Schiraldi, Glenn R. *The Post-Traumatic Stress Disorder Sourcebook: A Guide to Healing, Recovery, and Growth.* Los Angeles: Lowell House, 2000.

Sternberg, Esther M. *The Balance Within: The Science Connecting Health and Emotions.* New York: W. H. Freeman, 2001.

Summers, Caryn, ed. *Inspirations for Caregivers.* Mt. Shasta, Calif.: Commune-a-Key Publishing, 1993.

Utterback, Ann S. *Broadcast Voice Handbook: How to Polish Your On-Air Delivery.* 4th ed. Chicago: Bonus Books, 2005.

Wilson, Paul. *Instant Calm: Over 100 Easy-to-Use Techniques for Relaxing Mind and Body.* New York: Penguin Books, 1995.

WEB SITES

The sites *www.dartcenter.org* and *www.rsf.org* both contain very helpful material for any broadcaster covering crisis events. For general information about PTSD, go to *www.ncptsd.org.*

About Dr. Utterback

Dr. Ann S. Utterback is an internationally recognized Broadcast Voice Specialist and Stress Reduction Counselor with over thirty years of experience. She is a member of the American Academy of Experts in Traumatic Stress and has contributed to the World Congress of the International Critical Incident Stress Foundation. Her clients are from the major television networks, local affiliates, the Voice of America, CNN, FOX News, Bloomberg News, Reuters, and other television and radio stations throughout the United States and Canada. Dr. Utterback has traveled internationally giving workshops for stations, networks, and groups such as The Associated Press, American Forces Network in Europe, and the Alaska Press Club. She has been a regular speaker at the RTNDA International Conference where her lectures are rated among the highest. Following September 11, Dr. Utterback conducted workshops on dealing with crisis at WTOP Radio in Washington, D.C., the Voice of America, and other locations. Dr. Utterback's classic manual for on-air delivery, now in its fourth edition, *Broadcast Voice Handbook* is used extensively in newsrooms and university classrooms. Her book, co-authored with Professor Michelle McCoy, Kent State University, is *Sound and Look Professional on Television and the Internet: A Guide for Broadcasters and Executives.*

Dr. Utterback is based in the Washington, D.C. area. She sees clients in her office and offers telephone tape evaluations and consultations for clients who are not able to meet with her. Read more about her work and get an updated Voice Tip on her website: www.AVoiceDoc.com.

Other Acclaimed Broadcasting and Journalism Titles from Bonus Books

Writing Broadcast News, Rev. Ed.
Shorter, Sharper, Stronger - A Professional Handbook
By Mervin Block

ISBN: 1-56625-084-6 · $29.95 · 314 pp. · Cloth · 6" x 9"

A former staff writer for shows like NBC's "Today," Mervin Block knows what he is talking and writing about. He writes the way he talks, and he talks the way he writes. His book is readable—and enjoyable. And, for writers who apply its expert tips, profitable.

America's Right Turn: How Conservatives Used New and Alternative Media to Take Power
By Richard A. Viguerie and David Franke

ISBN: 1-56625-252-0 · $26.95 · 375 pp. · Cloth · 6" x 9"

Liberal media activists beware! Richard A. Viguerie and David Franke detail how conservatives—shut out by the liberal mass media of the 1950s and '60s—came to power by utilizing new and alternative media, and then created their own mass media.

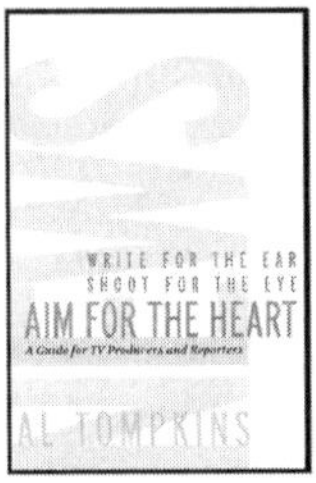

Aim for the Heart: Guide for TV Producers/Reporters
By Al Tompkins

ISBN: 1-56625-176-1 · $29.95 · 229 pp. · Cloth · 6" x 9"

The goal of this book is to help professional broadcast reporters, producers, and photojournalists tell stronger stories. It can serve as a guide for news managers whose job it is to train, coach, and inspire others. Veteran journalist Al Tompkins is Broadcast and Online Group Leader for the Poynter Institute for Media Studies.

Best Newspaper Writing 2004: The Nation's Best Journalism
Edited by Keith Woods

ISBN: 1-56625-234-2 · $19.95 · 396 pp. · Paper w/CD-ROM · 6" x 9"

Best Newspaper Writing 2004 celebrates the winners of the ASNE Distinguished Writing Awards. The book includes: the full text of the winning stories, interviews with the winners about their craft; study questions, and a companion CD-ROM containing all of the Community Service Photojournalism Award winners.

Broadcast Newswriting: The RTNDA Reference Guide
By Mervin Block

ISBN: 1-56625-017-X · $29.95 · 271 pp. · Cloth · 6" x 9"

Need a quick answer to a question — or problem — that pops up while you're writing news? The answer is in this book. It is the first specific reference guide for broadcast newswriting. A tool chest for professionals, it is full of tips on writing, language, and journalism.

Anchoring America: The Changing Face of Network News
By Jeff Alan with James M. Lane

ISBN: 1-56625-194-X · $24.95 · 426 pp. · Cloth · 6" x 9"

Anchoring America covers 19 anchors in 19 smart profiles that show the evolution of the anchoring job and reveal the character of the men and women who sat at the desk.

Broadcasting Through Crisis
How to Keep Going When Tragedy Hits
By Ann S. Utterback

ISBN: 1-56625-240-7 · $29.95 · 160 pp. · Cloth · 6" x 9"

The events of the past few years have produced some of the biggest crises our country has seen. Find out from the experts how to keep going when covering wars, terrorist events, weather emergencies and everyday tragedies.

A Few Marbles Left: A Close-Up Look at TV News in All Its Agonizing, Maddening Lunacy
By John Corcoran; Foreword by Mervin Block

ISBN: 1-56625-167-2 · $23.95 · 283 pp. · Cloth · 6" x 9"

A Few Marbles Left contains more than 75 essays about television news and other lunacies of the tube. John Corcoran serves up rants, recollections, behind-the-scenes bloopers and wry observations, as well as serious considerations on improving the craft.